Lecture Notes in Computer Science 11351

Commenced Publication in 1973
Founding and Former Series Editors:
Gerhard Goos, Juris Hartmanis, and Jan van Leeuwen

Editorial Board

More information about this series at http://www.springer.com/series/7412

Maria De Marsico · Gabriella Sanniti di Baja
Ana Fred (Eds.)

Pattern Recognition Applications and Methods

7th International Conference, ICPRAM 2018
Funchal, Madeira, Portugal, January 16–18, 2018
Revised Selected Papers

 Springer

Editors
Maria De Marsico
Computer Science
Sapienza University of Rome
Rome, Italy

Ana Fred
University of Lisbon
Lisbon, Portugal

Gabriella Sanniti di Baja
ICAR-CNR
Naples, Italy

ISSN 0302-9743   ISSN 1611-3349 (electronic)
Lecture Notes in Computer Science
ISBN 978-3-030-05498-4   ISBN 978-3-030-05499-1 (eBook)
https://doi.org/10.1007/978-3-030-05499-1

Library of Congress Control Number: 2018964896

LNCS Sublibrary: SL6 – Image Processing, Computer Vision, Pattern Recognition, and Graphics

This Springer imprint is published by the registered company Springer Nature Switzerland AG
The registered company address is: Gewerbestrasse 11, 6330 Cham, Switzerland

Preface

The present book includes the extended and revised versions of a set of selected papers from the 7th International Conference on Pattern Recognition Applications and Methods (ICPRAM 2018), held in Funchal, Madeira, Portugal, during January 16–18, 2018.

The ICPRAM series of conferences aims at becoming an important point of contact for researchers, engineers, and practitioners in the areas of pattern recognition, whose scope is continuously expanding. Both theoretical and application perspectives are jointly taken into account, aiming at an effective synergy with increasingly advanced implementations. The conference especially welcomes interdisciplinary research. The core of its program is intended to include theoretical studies yielding new insights in pattern recognition methods, as well as experimental validation and concrete application of pattern recognition techniques to real-world problems.

In this framework, ICPRAM 2018 received 102 submissions from 33 countries. The authors of the best conference papers were invited to contribute to this book by submitting the revised and extended versions of their conference papers. Candidate papers were selected by the ICPRAM 2018 chairs who based their choice on a number of criteria. These criteria included the scores and comments provided by the Program Committee reviewers, the assessment of the presentations by the session chairs, as well as the global view by program chairs of all the papers presented at ICPRAM 2018. The extensions were requested to have at least 30% innovative material including, e.g., further theoretical in-depth analysis and/or new experiments. The new papers underwent a two-round reviewing process, at the end of which the ten papers in this book were accepted.

We hope that this book will contribute to the understanding of relevant trends of current research in pattern recognition application and methods. Similarly to the conference organization, the ten papers in this book have been divided into the two main tracks: Applications and Methods. The seven papers dealing with methods are presented first, since they have a more general scope, and each of them may offer inspiration for different applications. Then the three papers presenting a wide variety of applications follow.

In "TIMIT and NTIMIT Phone Recognition Using Convolutional Neural Networks" by Cornelius Glackin, Julie Wall, Gérard Chollet, Nazim Dugan, and Nigel Cannings, an application of convolutional neural networks is presented for phone recognition. Both the TIMIT and NTIMIT speech corpora are employed and their phonetic transcriptions are used to label spectrogram segments in the training phase of the convolutional neural network. Phonetic rescoring is then performed to map each phone set to the smaller standard set.

In "Interactive Design Support for Architecture Projects During Early Phases Based on Recurrent Neural Networks" by Johannes Bayer, Syed Saqib Bukhari, and Andreas Dengel, the production of an architectural project is treated as an iterative design algorithm, in which high-level ideas and requirements are transformed into a specific building description. This process is usually carried out in a manual and labor-intensive manner, while in this paper a semi-automatic approach is presented to assist the developer by proposing suggestions for solving individual design steps automatically.

The paper "An Efficient Hashing Algorithm for NN Problem in HD Spaces" by Faraj Alhwarin, Alexander Ferrein, and Ingrid Scholl proposes a new hashing method to efficiently cope with nearest neighbor (NN) search in high-dimensional search spaces. In these spaces, the complexity grows exponentially with dimension and the data tends to show strong correlations between dimensions. The developed approach entails splitting the search space into subspaces based on a number of jointly independent and uniformly distributed circular random variables (CRVs), which are computed from the data points.

In "Stochastic Analysis of Time-Difference and Doppler Estimates for Audio Signals" by Gabrielle Flood, Anders Heyden, and Karl Aström, the pairwise comparison of sound and radio signals is considered to estimate the distance between two units that send and receive signals. Some methods are robust to noise and reverberation, but give limited precision, while sub-sample refinements are more sensitive to noise, but give higher precision when they are initialized close to the real translation. In this paper, stochastic models are presented that explain the precision limits of such sub-sample time-difference estimates.

The paper "Street2Fashion2Shop: Enabling Visual Search in Fashion E-commerce Using Studio Images," by Julia Lasserre, Christian Bracher and Roland Vollgraf, deals with visual search, in particular as regards the street-to-shop task of matching fashion items, displayed in everyday images, with similar articles. Street2Fashion2Shop is a pipeline architecture that stacks Studio2Fashion, a segmentation model responsible for eliminating the background in a street image, with Fashion2Shop, an improved model matching the remaining foreground image with "title images." The latter are front views of fashion articles on a white background. Both segmentation and product matching rely on deep convolutional neural networks. The pipeline allows for circumventing the lack of quality annotated wild data by leveraging specific data sets at all steps.

In the paper "Earth Mover's Distance Between Rooted Labeled Unordered Trees Formulated from Complete Subtrees" by Taiga Kawaguchi, Takuya Yoshino and Kouichi Hirata, earth mover's distances (EMDs) are introduced for rooted labeled trees formulated from complete subtrees. The EMDs are shown to be metrics and are computed in $O(n3\{\backslash\}\log n)$ time, where n is the maximum number of nodes in two given trees.

The paper "CNN-Based Deep Spatial Pyramid Match Kernel for Classification of Varying Size Images" by Shikha Gupta, Manjush Mangal, Akshay Mathew, Dileep Aroor Dinesh, Arnav Bhavsar, and Veena Thenkanidiyoor addresses the issues of handling varying size images in convolutional neural networks (CNNs). Two approaches are considered. The first approach explores deep spatial pyramid match kernel (DSPMK) to compute a matching score between two varying size sets of activation maps. In the second approach, spatial pyramid pooling (SPP) layer in CNN architectures is used to remove the fixed-length constraint, and to allow the original varying size image as input to train and fine-tune the CNN for different datasets.

In "Detection and Classification of Faulty Weft Threads Using Both Feature-Based and Deep Convolutional Machine Learning Methods" by Marcin Kopaczka, Marco Saggiomo, Moritz Güttler, Kevin Kielholz, and Dorit Merhof, a novel computer vision approach is presented to automatically detect faulty weft threads on air-jet weaving machines. The system consists of a camera array for image acquisition and a classification pipeline, where different image processing and machine learning methods are used to precisely localize and classify defects.

Video activity recognition using support vector machines (SVMs) is treated in "Video Activity Recognition Using Sequence Kernel-Based Support Vector Machines" by Sony S. Allappa, Veena Thenkanidiyoor, and Dileep Aroor Dinesh. Videos include sequences of sub-activities, where sub-activities correspond to video segments. Segments are encoded by means of feature vectors and each video is represented as sequences of feature vectors. GMM-based encoding schemes and bag-of-visual-word vector representation are considered to encode video segments. Building the SVM-based activity recognizer requires suitable kernels matching the sequences of feature vectors. To this purpose, time flexible kernel, segment level pyramid match kernel, segment level probability sequence kernel, and segment level Fisher kernel are employed.

In "Visual Cryptography for Detecting Hidden Targets by Small-Scale Robots" by Danilo Avola, Luigi Cinque, Gian Luca Foresti and Daniele Pannone, a vision-based system is presented to find encrypted targets in unknown environments by using small-scale robots and visual cryptography. The robots acquire the scene by a standard RGB camera and use a visual cryptography-based technique to encrypt the data. Encrypted data are subsequently sent to a server that will decrypt and analyze it for searching target objects or tactic positions. The experiments have been performed by using two robots, i.e., a small-scale rover in indoor environments and a small-scale unmanned aerial vehicle (UAV) in outdoor environments, to show the effectiveness of the proposed system.

We would like to close this preface by thanking all the authors for their contributions and also the reviewers, who helped to ensure the quality of this publication. Pattern recognition is a continuously evolving and multifaceted discipline, and it is hard to account for all possible applications, but we hope that the papers in this book can provide a significant sample or present research to encourage new achievements.

January 2018

Maria De Marsico
Gabriella Sanniti di Baja
Ana Fred

Organization

Conference Chair

Ana Fred   Instituto de Telecomunicações and Instituto Superior Técnico - Lisbon University, Portugal

Program Co-chairs

Maria De Marsico   Sapienza Università di Roma, Italy
Gabriella Sanniti di Baja   Italian National Research Council CNR, Italy

Program Committee

Andrea Abate   University of Salerno, Italy
Ashraf AbdelRaouf   Misr International University MIU, Egypt
Zeina Abu-Aisheh   Université de Tours, France
Gady Agam   Illinois Institute of Technology, USA
Lale Akarun   Bogazici University, Turkey
Adib Akl   Holy Spirit University of Kaslik, Lebanon
Mayer Aladjem   Ben-Gurion University of the Negev, Israel
Javad Alirezaie   Ryerson University, Canada
Kevin Bailly   Pierre and Marie Curie University (UPMC), France
Emili Balaguer-Ballester   Bournemouth University, UK
Vineeth Balasubramanian   Indian Institute of Technology, India
Enrique Ballester   Università degli Studi di Milano, Italy
Imen Ben Cheikh   University of Tunis, Tunisia
Mohammed Bennamoun   The University of Western Australia, Australia
Stefano Berretti   University of Florence, Italy
Monica Bianchini   University of Siena, Italy
Michael Biehl   University of Groningen, The Netherlands
Battista Biggio   University of Cagliari, Italy
Isabelle Bloch   Télécom ParisTech, Université Paris-Saclay, France
Andrea Bottino   Politecnico di Torino, Italy
Nizar Bouguila   Concordia University, Canada
Francesca Bovolo   Fondazione Bruno Kessler, Italy
Paula Brito   Universidade do Porto, Portugal
Javier Calpe   Universitat de València, Spain
Francesco Camastra   University of Naples Parthenope, Italy
Virginio Cantoni   Università di Pavia, Italy
Michelangelo Ceci   University of Bari, Italy
Jocelyn Chanussot   Grenoble Institute of Technology Institut Polytechnique de Grenoble, France

Antonio-José Sánchez-Salmerón	Universitat Politecnica de Valencia, Spain
Carlo Sansone	University of Naples Federico II, Italy
K. C. Santosh	The University of South Dakota, USA
Michele Scarpiniti	Sapienza University of Rome, Italy
Paul Scheunders	University of Antwerp, Belgium
Leizer Schnitman	Universidade Federal da Bahia, Brazil
Friedhelm Schwenker	University of Ulm, Germany
Ishwar Sethi	Oakland University, USA
Linda Shapiro	University of Washington, USA
Lauro Snidaro	Università degli Studi di Udine, Italy
Humberto Sossa	Instituto Politécnico Nacional-CIC, Mexico
Tania Stathaki	Imperial College London, UK
Mu-Chun Su	National Central University, Taiwan
Johan Suykens	KU Leuven, Belgium
Eulalia Szmidt	Systems Research Institute Polish Academy of Sciences, Poland
Vahid Tabar	Allameh Tabataba'i University, Iran, Islamic Republic of
Xiaoyang Tan	Nanjing University of Aeronautics and Astronautics, China
Andrea Torsello	Università Ca'Foscari Venezia, Italy
Godfried Toussaint	New York University Abu Dhabi, UAE
Rosa Valdovinos Rosas	Universidad Autonoma del Estado de Mexico, Mexico
Ernest Valveny	Universitat Autònoma de Barcelona, Spain
Antanas Verikas	Intelligent Systems Laboratory, Halmstad University, Sweden
Markus Vincze	Technische Universität Wien, Austria
Panayiotis Vlamos	Ionian University, Greece
Asmir Vodencarevic	Siemens Healthcare GmbH, Germany
Laurent Wendling	LIPADE, France
Slawomir Wierzchon	Polish Academy of Sciences, Poland
Shengkun Xie	Ryerson University, Canada
Jing-Hao Xue	University College London, UK
Chan-Yun Yang	National Taipei University, Taiwan
Slawomir Zadrozny	Polish Academy of Sciences, Poland
Pavel Zemcik	Brno University of Technology, Czech Republic
Jing Zhao	ECNU, China
Fei Zhou	Tsinghua University, China
Huiyu Zhou	Queen's University Belfast, UK

Additional Reviewers

Shuo Chen	Shenzhen Institutes of Advanced Technology, China
Ting-Li Chen	Academia Sinica, Taiwan
Ivan Duran-Diaz	University of Seville, Spain

Daniel Gómez-Vergel   Universidad Europea de Madrid, Spain
Ali Hamou   The University of Western Ontario, Canada
Eddy Ihou   Concordia University, Canada
Krishna Kakkirala   Tata Consultancy Services, India
Jose López-López   Universidad Europea de Madrid, Spain
Rui Zhu   University of Kent, UK

Invited Speakers

Rita Cucchiara   University of Modena and Reggio Emilia, Italy
Edwin Hancock   York University, UK
Xiaoyi Jiang   University of Münster, Germany
Alfred Bruckstein   Technion, Israel

Contents

Theory and Methods

Applications

Theory and Methods

Street2Fashion2Shop: Enabling Visual Search in Fashion e-Commerce Using Studio Images

Julia Lasserre$^{(\boxtimes)}$, Christian Bracher, and Roland Vollgraf

Zalando Research, Mühlenstr. 25, 10243 Berlin, Germany
{julia.lasserre,christian.bracher,roland.vollgraf}@zalando.de

Abstract. Visual search, in particular the *street-to-shop* task of matching fashion items displayed in everyday images with similar articles, is a challenging and commercially important task in computer vision. Building on our successful Studio2Shop model [20], we report results on *Street2Fashion2Shop*, a pipeline architecture that stacks Studio2Fashion, a segmentation model responsible for eliminating the background in a street image, with Fashion2Shop, an improved model matching the remaining foreground image with "title images", front views of fashion articles on a white background. Both segmentation and product matching rely on deep convolutional neural networks. The pipeline allows us to circumvent the lack of quality annotated wild data by leveraging specific data sets at all steps. We show that the use of fashion-specific training data leads to superior performance of the segmentation model. Studio2Shop built its performance on FashionDNA, an in-house product representation trained on the rich, professionally curated Zalando catalogue. Our study presents a substantially improved version of FashionDNA that boosts the accuracy of the matching model. Results on external datasets confirm the viability of our approach.

Keywords: Visual search · Computer vision · Deep learning
Product matching · Fashion

1 Introduction

Online fashion sales are already worth billions of dollars, and keep growing rapidly. An ultraconnected millenial generation spends large amounts of time on social networks such as Instagram. As a result, the quest for fashion inspiration is being deeply transformed. Instead of browsing shop displays, they take screenshots of images with fashion favourites and expect to be able to search for them on their smartphone. Hence, "street-to-shop,", the task of retrieving products that are similar to garments depicted in a "wild" image of varying quality and non-uniform background, is a critical capability for online stores. More and more of them now offer visual search features, Asos in the UK being one of the latest examples [29].

© Springer Nature Switzerland AG 2019
M. De Marsico et al. (Eds.): ICPRAM 2018, LNCS 11351, pp. 3–26, 2019.
https://doi.org/10.1007/978-3-030-05499-1_1

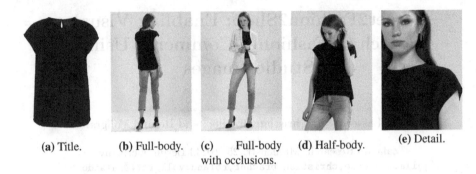

(a) Title.   (b) Full-body.   (c)   Full-body   (d) Half-body.   (e) Detail.
                                    with occlusions.

Fig. 1. Typical images from our catalogue for a black t-shirt.

The surge of computational fashion is reflected by a growing number of academic contributions to the topic, often in partnership with online shops. A lot of effort focuses on representation learning for apparel [1,3,11,28], where fashion items are encoded as feature vectors in a high-dimensional map that is implicitly structured by their properties. Many efforts are also spent on building *taggers*, models that predict attributes of products such as yellow, v-neck and long sleeves [2,4,5]. Another line of work is clothing parsing [7,31,34,35], which consists in finding various apparel categories in the image, pixel-wise. Magic mirrors are also popular [6,13,18,23,30,36], and fashion synthesis is gaining attention [14,40]. Large companies also contribute, for example Pinterest [16] or eBay [37], even if the actual architectures or datasets are not shared.

Research on the street-to-shop task started with [24], and has grown into a large body of literature. Early studies [8,17,24,33,34] rely on classical computer vision and use tools such as body part detection and hand-crafted features. A few years ago, deep learning has become the norm, however. Recent studies typically share a common principle: Feature representations are learnt for both query images and products via attribute classification, then a ranking loss is employed which encourages matching pairs to have higher scores [12,15,25,27,28,32]. Often, siamese architectures are used to relate the two domains [12,32], but some studies do not make a dichotomy between query and product images and employ a single branch [25].

As the leading European online fashion platform, which has been active for ten years, Zalando has accumulated a huge catalogue containing millions of items, each of which is described by a set of studio images (Fig. 1) and a set of attributes. This data is leveraged throughout the company to provide internal and customer-facing services. At Zalando Research we continuously develop the representation learning framework FashionDNA [3,11]. It serves both as a practical numerical interface for visual information, and as a research tool for fashion computation. In particular, it provides a compact product representation in our visual search model Studio2Shop [20]. Studio2Shop matches studio model images (with uniform background, as in Figs. 1b–d) with fashion articles. It works well, but the restriction to studio model images limits its practical application, and we aim to extend it to real-life (wild) images with various types of backgrounds and different illuminations.

To extend Studio2Shop to a true street-to-shop model, we can think of two obvious approaches:

- Replace the studio images with "wild" image data, and retrain,
- build a pipeline that first performs image segmentation, *i.e.* separates the background from the person shown, then matches the latter image to fashion products.

While the first approach is direct, more elegant, and likely faster at test-time, obtaining training data with appropriate annotations in sufficient numbers and quality is a time-consuming and costly endeavour, as there is a shortage of publicly available data, and images would have to be annotated manually to retain the quality of studio image annotations. The second approach allows us to retain our large catalogue dataset as primary source to learn image–product matching. Moreover, a body of ground truth data for the segmentation model is available for academic purposes, and the image segmentation task can easily be crowd-sourced for commercial purposes. Recent models [9,26] are mature enough to quickly adapt working techniques.

Here, we follow the second approach and show that it is promising. Our main contribution is three-fold:

- We extend Studio2Shop to wild images,
- we keep all the former contributions of Studio2Shop: (a) to require no user input, (b) to naturally handle all fashion categories, and (c) to show that static product representations are effective.
- we achieve reasonable results on external datasets without fine-tuning our model.

We now proceed with an overview of our model pipeline (Sect. 2), and describe its components in more detail in Sects. 3–5. Section 6 shows our results on external datasets. Finally, Sect. 7 concludes the study.

2 Street2Fashion2Shop: The Pipeline

Figure 2 shows the working pipeline of Street2Fashion2Shop. On the top row, the wild query image is segmented by Street2Fashion in order to discard the background and keep the fashion only. On the bottom row, our product representation FashionDNA is applied to the title images of the set of products we should retrieve from (the assortment), in order to obtain static feature vectors. Finally the segmented query image and the product feature vectors (the fDNAs) form the input of the product matching model Fashion2Shop which gives a score to each product conditioned on the query image. Finally products are ranked by relevance (decreasing score) and the first ones are displayed.

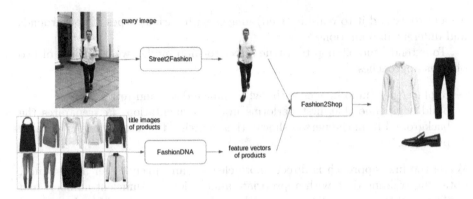

Fig. 2. The pipeline of Street2Fashion2Shop. The query image (top row) is segmented by Street2Fashion, while FashionDNA is run on the title images of the products in the assortment (bottom row) to obtain static feature vectors. The result of these two operations forms the input of Fasion2Shop which handles the product matching.

3   Fashion DNA "1.1" (fDNA1.1)

FashionDNA (fDNA) is the pipeline component that provides a fixed numerical article encoding. For convenience, we denote the previous version used in [20] as FashionDNA 1.0, and now describe the current module, FashionDNA 1.1.

The wealth of curated fashion article data owned by Zalando enables us to find meaningful item representations using deep learning methods. For the present work, we employ article embeddings of dimension $d = 128$, as opposed to 1536 previously, that are extracted as hidden activations of a fully convolutional neural network with residual layers. Images of products presented on a white background ("title" images) serve as input, and the network is tasked to predict thousands of binary labels (tags) describing the article.

3.1   Data

FashionDNA is meant to be of general purpose for Zalando and uses a very large set of ~2.9 M items. For these products, we retrieve high-resolution title images, of the kind shown in Fig. 1a, typically in an upright 762×1100 pixel format, although exact dimensions can vary. These images are downscaled to the biggest size that would fit a canvas of dimension 165×245, preserving their aspect ratio, and placed in its center. We pad images with white background as necessary.

As labels, we use curated item attributes that are assigned by fashion experts at the time of image production. Attributes include general properties (like brand and *silhouette*, describing the functional article category), as well as tags assigned for specific product groups (e. g., neckline tags for the *shirt* silhouette). They possess wide variations in frequency, consistency, and visibility on images. We roll out all existing attribute options into a sequence of boolean labels of equal status. For fDNA 1.1, we obtain 11,668 such labels, as opposed to 6,092 for

FashionDNA 1.0. These 11,668 labels include >6k distinct brands, and 5–25 of them typically assigned per article. Their population has long-tail characteristics, where most labels occur only rarely. The Shannon entropy quantifies the information contained in the label distribution alone, disregarding the images:

$$\mathcal{H}_S = -\sum_{\lambda=1}^{L} q_\lambda \log q_\lambda \approx 53.5 \,, \tag{1}$$

where q_λ is the frequency of label λ among items. This entropy serves as a reference point in training our image-to-label classification model.

3.2 The Model

Architecture. For fDNA 1.1, we employ a fully convolutional residual network [10] with batch normalization and a bottleneck layer for fDNA extraction that was purpose-built and trained using Tensorflow. For architecture details we refer to Tables 1a and 1b. We note that the current model is much deeper than the network used in [20] which rested on the AlexNet model [19], and provides superior quality embeddings with fewer parameters and fewer dimensions (128 instead of 1536).

Loss Function. We train our network to minimize the cumulative cross-entropy loss, averaged over training set items. To estimate it, we compare the predicted probability $p_{k\lambda}$ for the kth article to carry label λ in a minibatch of $K = 128$ with the corresponding ground truth $y_{k\lambda} \in \{0, 1\}$, and sum over all labels:

$$\mathcal{L}_{\text{mini}} = -\frac{1}{K}\sum_{k=1}^{K}\sum_{\lambda=1}^{L} \left(y_{k\lambda} \log p_{k\lambda} + (1 - y_{k\lambda}) \log(1 - p_{k\lambda})\right) . \tag{2}$$

Note that all binary labels carry equal weight in our model.

Training Procedure. We split the articles into training ($\sim$2.8M items, 96%) and test sets ($\sim$110k items). Unlike some embedding models for fashion images [1,28] that foot on pre-trained, open-sourced models, we train our network from scratch. We use Glorot initialization for all layers, save for the logistic layer bias which we pre-populate with the inverse sigmoid function $\sigma^{-1}(q_\lambda)$ of the observed label frequency, so that the initial network loss (2) matches the Shannon entropy $\mathcal{H}_S$ (1). The model is trained for 11 epochs using AdaGrad as an optimiser.

3.3 Results

We reach a loss of $\mathcal{L} \approx 19.5$, as opposed to the initial 53.5. In comparison, FashionDNA 1.0 only reached 22.5. We use test data to assess FashionDNA 1.1 on the task of predicting some product categories of interest to this manuscript: blazer, dress, pullover, shirt, skirt, t-shirt/top, trouser. We use as metric the

Table 1. Architecture of FashionDNA 1.1.

name	type	activation	output size	# params
input image	InputLayer		165x245x3	0
preprocess	Invert + Scale		165x245x3	1
encoding	Convolution 1		79x119x64	15,552
	Residual Block × 1	ReLU	79x119x64	73,984
	BatchNorm or Scale	ReLU	79x119x64	128
	Convolution 2		39x59x128	73,728
	Residual Block × 2	ReLU	39x59x128	590,848
	BatchNorm or Scale	ReLU	39x59x128	256
	Convolution 3		19x29x256	294,912
	Residual Block × 3	ReLU	19x29x256	3,540,992
	BatchNorm or Scale	ReLU	19x29x256	512
	Convolution 4		9x14x512	1,179,648
	Residual Block × 2	ReLU	9x14x512	9,441,280
	BatchNorm or Scale	ReLU	9x14x512	1,024
	Spatial AvgPool		512	0
	Dense		128	65,536
FashionDNA	Bottleneck		**128**	
output layer	Dense	Sigmoid	11,668	1,505,172
label prob	Probabilities		11,668	0
				16,783,573

(a) Global architecture of FashionDNA 1.1. The input is a minibatch of 128 title images, the output is a prediction for each attribute based on a sigmoid function.

name	type	activation	output size	# params
input	InputLayer		$D_1 \times D_2 \times C$	0
residual block	BatchNorm or Scale	ReLU		$2C$
	Convolution 1			$9C^2$
	BatchNorm or Scale	ReLU		$2C$
	Convolution 2			$9C^2$
	Add Input Layer			0
output	OutputLayer		$D_1 \times D_2 \times C$	0
				$2C(9C + 2)$

(b) Structure of a residual block. The input tensor undergoes a sequence of batch normalization (or shifting & scaling during fine-tuning and inference) and dimension-preserving convolution with a 3×3 kernel twice. The result is added element-wise to the input tensor.

area under the ROC curve, which should be 0.5 for a random guess, and 1 for a perfect model. For all the categories of interest, the area under the curve was above 0.99. These results are very encouraging, as fDNA1.1 is clearly able to encode much of the information needed for visual search.

Evaluating the ability of a fashion item embedding to create meaningful neighbourhoods is conceptually difficult, as the notion of product similarity is in the eye of the observer. Our model has the capacity to use many different modes of similarity (like silhouette, color, function, material, etc.) simultaneously to achieve the match between two quite different presentations of the same article. Sampling nearest neighbors in fDNA space hints indeed at such multi-modal behavior (Fig. 3): A pair of snow pants is mostly matched by function, a dress based on color and embellishment (lace inserts).

4   Street2Fashion (segmentation)

Street2Fashion is the component of the pipeline responsible for segmenting out the background, *i.e.* everything which is not people or fashion items.

4.1   Data

Street2Fashion is trained using a mixture of publicly available data and data from our shop (both from the catalogue and from street shots). Our complete dataset is made of:

- 19554 fashion images mostly from Chictopia10K [21,22] and Fashionista [22, 35] where various categories of garments, together with hair and skin, are segmented separately. We do not use the category-specific segmentations, instead we combine them and treat them all as foreground.
- 200 studio model images from our catalogue, segmented by us.
- 90 model images from our street shots, segmented by us.

We cannot release our own images, however much of the results can be reproduced using the public Chictopia and Fashionista datasets.

In addition, images are slightly altered to increase the robustness of the segmentation model. As described in Appendix Sect. A, the set of transformations is {"none", "translate", "rotate", "zoom in", "zoom out", "blur", "noise", "darken", "brighten"}. Additionally, this set is doubled by applying a horizontal flip

Fig. 3. Nearest neighbors of sample articles (left column) among a test set of 50k clothing items in the FashionDNA embedding space.

("fliph") to the image first. All images undergo all transformations, with the parameters of the transformation each time randomly sampled from an acceptable range, thereby inflating the dataset by a factor of 18. Images are then resized to 285 × 189. Typical images from each source are shown, after transformation, in Fig. 5 in Sect. 4.3.

4.2   Model

Architecture. Our segmentation model follows the idea of a U-net architecture [26], as given in Table 2. The input is an image of size (285, 189, 3) whose values are divided by 255 to lie between 0 and 1. The output is an image where the person has been identified and the original background is replaced by white. The values also lie between 0 and 1.

Table 2. Architecture of Street2Fashion.

name	type	activation	output size	#params
input image	inputLayer		285×189×3	0
encoding	Convolution 1	ReLU	94×62×32	3,488
	Convolution 2	ReLU	46×30×128	65,664
	Convolution 3	ReLU	22×14×256	524,544
	Convolution 4	ReLU	10×6×256	1,048,832
	Convolution	ReLU	4×2×256	1,048,832
	Flatten		2048	0
	Dense		1024	2,098,176
decoding	Dense		2048	2,099,200
	Reshape		4×2×256	0
	Transposed convolution		10×6×256	1,048,832
	Concatenate (with Conv 4)		10×6×512	0
	Transposed convolution		22×14×256	2,097,408
	Concatenate (with Conv 3)		22×14×512	0
	Transposed convolution		46×30×128	1,048,704
	Concatenate (with Conv 2)		46×30×256	0
	Transposed convolution		94×62×32	131,104
mask	Transposed convolution	Sigmoid	285×189×1	1,153
rgb mask			285×189×3	0
foreground	input image × mask		285×189×3	0
background	1 - mask		285×189×3	0
output image	foreground + background		285×189×3	0
				11,215,937

Backward Pass. The labels or targets are the original images where the background has been replaced with white pixels, and the loss is the mean-square-error between the corresponding pixels:

$$\mathcal{L} = \frac{1}{N} \sum_{i=1}^{N} \left(\frac{1}{J} \sum_{j=1}^{J} (v_{ij} - l_{ij})^2 \right)$$

where N is the number of images in the batch, J the number of elements per image ($285 \times 189 \times 3 = 161595$ elements), v_{ij} the predicted value for the jth element of the ith image, and l_{ij} the ground-truth value of that element. We considered a binary cross-entropy loss directly on the mask itself, but that changed neither the global performance nor the aspect of the segmentation.

4.3 Results

Experimental Set-up. We randomly split the dataset, keeping 80% for training, and setting the rest aside for testing. The optimiser is Adam, with a learning rate of 0.0001 and other parameters set to default, the batch size is 64. Our internal images, due to their much lower representation, are upweighted by a factor of 10 in the loss. Performance is measured using two metrics: mean-square-error (mse) and accuracy. Both metrics are computed at the pixel level using the soft mask predicted by the model of interest. The mse compares the true segmented image to the soft segmented image obtained from the soft mask:

$$\text{mse(image)} = \frac{1}{J} \sum_{j=1}^{J} (v_j - l_j)^2$$

where J is the number of elements per image ($285 \times 189 \times 3 = 161595$ elements), v_j the predicted value for the jth element of the image, and l_j the ground-truth value of that element. The accuracy compares the true mask to the hard mask obtained from the soft mask (the soft values higher than or equal to 0.5 become 1, the others 0):

$$\text{accuracy(image)} = \frac{1}{J} \sum_{j=1}^{J} \mathbb{1}_{(q_j = m_j)}$$

where q_j is the predicted hard value for the jth element of the mask, and m_j the ground-truth value of that element. We want low mse and high accuracy.

Comparison with Mask-RCNN [9]. We run the publicly available keras implementation of Mask-RCNN, a state-of-the-art model for multiclass segmentation, on our test images. We keep the image size and the set of images the same as above. For Mask-RCNN, we consider as fashion the categories {person, backpack, umbrella, handbag, tie, suitcase} as they would all be labelled as foreground in our ground truth data. Figure 4 shows the distributions of these metrics at the image level (averaged over all pixels in an image), in green for

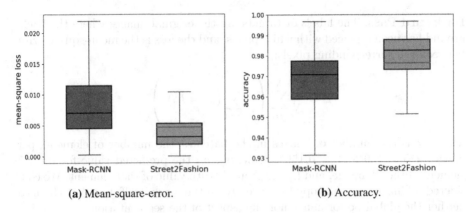

(a) Mean-square-error.   (b) Accuracy.

Fig. 4. Distribution of metrics at the image level, in green for Mask-RCNN [9], in orange for Street2Fashion. In each boxplot, the black horizontal middle line represents the median, while the red line represents the mean. (Color figure online)

Mask-RCNN, in orange for our model. In each boxplot, the black horizontal middle line represents the median, while the red line represents the mean.

Street2Fashion performs significantly better on these fashion images than Mask-RCNN, despite being much simpler, however this statement should be interpreted very carefully. Street2Fashion has trained on the same distribution of images as the test images, it is tuned to solve this task, but it can only process people and fashion accessories such as umbrellas or bags, it only knows these categories as "foreground" and it does not generalise to crowds. In contrast, Mask-RCNN is more powerful as it can do many more categories and can also deal with several instances of each category. However our use case is specifically about such fashion images, which implies that we do not need to be general or to deal with multiple instances. We could fine-tune Mask-RCNN to our task and benefit from its flexibility, however, because it is much simpler, Street2Fashion is also much faster, which is important at test time. For a batch size of 16 images, without any image pre-processing or results post-processing involved, applying the keras method predict_on_batch on the same GPU takes 0.04 s for Street2Fashion, 5.25 s for Mask-RCNN (these numbers averaged over 500 batches and are only a rough estimation).

Examples of Segmentations. Figure 5 shows random examples of test segmentations for the four image sources used in training. Generally the results are encouraging, and even if they are not always perfect, they are good enough to retain most of the clothing while discarding most of the background.

5 Fashion2Shop (Product Matching)

Fashion2Shop is the component of the pipeline responsible for matching query images with white backgrounds to relevant products. It uses the same dataset and the same architecture as Studio2Shop [20], however there are a few key differences:

- FashionDNA 1.1 is of better quality than FashionDNA 1.0 [20], see Sect. 3. In addition, its dimensionality is much smaller (128 against 1536), so unlike in [20] no dimensionality reduction method is needed.
- The training uses model images segmented by Street2Fashion, see Sect. 4.
- In order to be more robust to unknown types of images, the image transformations described in Appendix Sect. A are used during training. The dataset is not inflated but a transformation and its parameters are sampled randomly for each image of each batch.

5.1   Data [20]

Catalogue Images. Zalando has a catalogue rich of millions of fashion products, each of which is described by a set of studio images. For historical and practical reasons, for the visual search task, we focus on 7 types of articles: dress, blazer, pullover, t-shirt/top, shirt, skirt and trousers, however the approach put forward is not formally restricted to a number of categories and can be extended in a straightforward manner.

Figure 1 shows an example of our catalogue images for a black t-shirt. All our images follow a standard format (though the standard format may change over time) of two main types:

- The title image, which shows a front view of the article on a (invisible) hanger, see Fig. 1a.
- (studio) model images, which show various views of the article being worn by a model with a clean background, and can have three different formats:
  - Full-body: these images show the full person and usually display more than one article, see Fig. 1b. There can be occlusions, see Fig. 1c.
  - Half-body: these images focus on the article, see Fig. 1b.
  - Detail: these images focus on a detail of the article (say the zipper of a cardigan) and show fewt to no body parts, see Fig. 1e. These "detail" images are challenging because it is often very hard to tell even for humans what kind of product it is even supposed to be.

As in [20], the product is represented by FashionDNA, a numerical vector learnt separately on title images. Unlike [20] however, we do not need to apply PCA to reduce dimensionality as fDNA1.1 vectors are already of dimension 128, which allows us not to lose any information. Indeed, as stressed in Sect. 3, the product representation in this manuscript is quite different from [20], both in terms of model and quality. Having this representation is a strong asset, but in order to assess the generalising capacity of our model properly, title images should not be used as query images. This can change for a model used in production however, and should not constitute a limitation.

Model images are all considered query images. We take the first 4 model images per article, or all of them if there are fewer than 4. This is enough

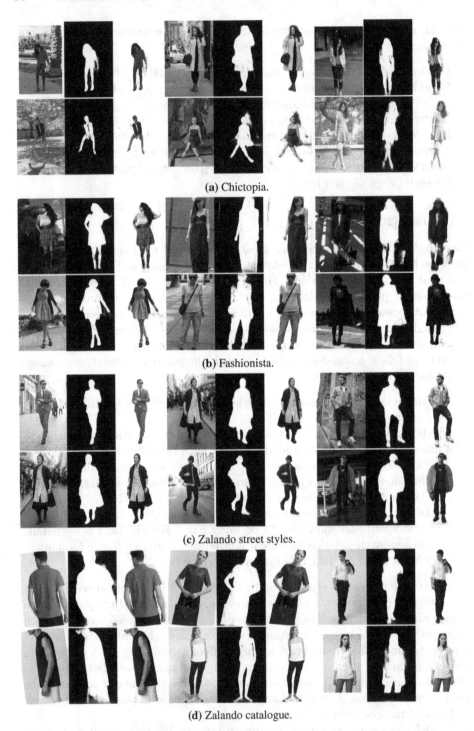

(a) Chictopia.

(b) Fashionista.

(c) Zalando street styles.

(d) Zalando catalogue.

Fig. 5. Examples of segmentation results on test images.

to capture most full/half-body images and discard as many detail images as possible. In total, in this manuscript, we use 246,961 products and 957,535 model images for training, 50000 products and 20000 images for testing.

Annotations. In visual search we are interested in retrieving products relevant to a query image. For training, we need examples of such relevant products, which we easily get for the studio model images from our catalogue. Several products are usually visible on model images, however usually only the product for which the image was shot is entered into the system. As a result, most of our model images are annotated with this one product of interest only.

For a (increasing) minority of the full-body images however, more than one product are entered into the system. In our dataset, this minority is made of 126713 images (about 13%). This subset of well annotated images allows us to find several types of products in most full-body images. The maximum number is 4, but most images (about 80%) only have 1. Note that we only keep the annotations that belong to the 7 types of articles of interest. This multiple annotations are precious but not perfect. A lot of the extra products in these images overlap, *i.e.* some products are re-used for different shots, some more than 100 times or even 700 times. A typical case is a pair of black skinny jeans that fits virtually every t-shirt/top/blouse. Models will change their top for different shots but will keep the jeans for efficiency. Another example is a plain t-shirt or tank top, often worn under the shirt and therefore not visible, which creates inconsistent annotations.

Table 3. Architecture of Model2Shop.

submodule	name	type	activation	output size	#params
query-feature	query_input	InputLayer		224×155×3	0
		VGG16		7×4×512	14,714,688
		Flatten		14436	0
		Dense	ReLU	2048	29,362,176
		Dropout(0.5)		2048	0
		Dense	ReLU	2048	4,196,352
		Dropout(0.5)		2048	0
	query_fDNA	Dense		128	262,272
query-article-matching	article_fDNA	InputLayer		128	0
	query_fDNA	InputLayer		128	0
		Concatenate		256	0
		BatchNorm		256	512
		Dense	ReLU	256	65792
		Dense	ReLU	256	65792
	match probability	Dense	Sigmoid	1	257
					48,667,841

5.2 Model

Architecture [20]**.** The architecture is the same as in [20] and given in Table 3. It has a left leg (or query leg), whose input is referred to as query_input in the table, and a short static right leg (or product leg), whose input is referred to as article_fDNA in the table.

Training. The training procedure is as described in [20]. The query leg is fed with model images, the product leg with its fDNA. Model images are resized to 224×155, transformed randomly using one of the transformations in Appendix Sect. A and segmented with Street2Fashion. Each query image is run against 50 products: we first take all the products annotated, they will constitute our positive articles, and we complete the 50 slots using randomly sampled products, which will constitute our negative articles. The positives are always the same but the negatives are sampled randomly for each batch and for each epoch. For a mini-batch size of N, we therefore get $50 \times N$ matches x, some positive with label $y = 1$, some negative with label $y = 0$. The optimiser is Adam, with a learning rate of 0.0001 and other parameters set to default, the batch size is 64.

Backward Pass. We use a cross-entropy loss:

$$\mathcal{L} = \sum_{i=1}^{N} \sum_{j=1}^{50} \left(y_{ij} \log p\left(I_i, x_{ij}\right) + \left(1 - y_{ij}\right) \log\left(1 - p(I_i, x_{ij})\right) \right)$$

where I_i is the ith model image, x_{ij} the jth product for that image, $p\left(I_i, x_{ij}\right)$ the probability given by the model for the match between image I_i and product x_{ij}, and y_{ij} the actual label of that match (1 or 0).

Testing. At test time, each test query image is matched against a gallery of previously unseen products, and these products are then ranked by decreasing matching probability. The plots in this manuscript typically show the top 50 suggested products for each image.

5.3   Competing Architectures [20]

In [20], Studio2Shop is tested alongside different variants of its architecture, summarised below. For more detailed information, please refer to the original manuscript.

- fDNA1.0-ranking-loss (formerly referred to as fDNA-ranking-loss) uses a simple dot product to order the (query image, product) pairs, and a sigmoid ranking loss.
- fDNA1.0-linear uses (formerly referred to as fDNA-linear) a simple dot product to order the (query image, product) pairs, and a cross-entropy loss.
- Studio2Shop with fc14 (formerly referred to as fc14-non-linear) has the same architecture and loss as Studio2Shop but uses VGG features for products instead of fDNA1.0.
- Studio2Shop with 128floats (formerly referred to as 128floats-non-linear) has the same architecture and loss as Studio2Shop but uses 128-floats features for products instead of fDNA1.0.

5.4   Results

The data and the architecture of Model2Shop are the same as in [20], so we can assess FashionDNA 1.1 and Street2Fashion.

Experimental Set-up. For ease of comparison, we follow the same procedure as described in [20], and run tests on 20000 randomly sampled unseen test query images against 50000 unseen test products. For each test query image, all 50000 possible (image, product) pairs are submitted to the model and are ranked by decreasing score.

Performance. Table 4 shows performance results on the various models and compare them to [20]. We assess our models using top-k retrieval, which measures the proportion of query images for which a correct product is found in the top k suggestions. The top-1% measure (which here means top-500) is given for easier comparisons in case the gallery of a different dataset has a different size. The average metric gives the average index of retrieval: an average of 5 means that a correct product is found on average at position 5. Because the distribution of retrieval indices is typically heavy-tailed, we add the median metric which gives the median index of retrieval: a median of 5 means that for half the images, a correct product is found at position 5. All our models are assessed on the exact same (query image, product) pairs.

Table 4. Results of the retrieval test using 20000 query images against 50000 Zalando articles. Top-k indicates the proportion of query images for which the correct article was found at position k or below. Average and median refer respectively to the average and median position at which an article is retrieved. The best performance is shown in bold.

	top-1	top-5	top-10	top-20	top-50	top-1%	average	median
fDNA1.0-ranking-loss [20]	0.091	0.263	0.372	0.494	0.660	0.933	177	20
fDNA1.0-linear [20]	0.121	0.314	0.423	0.547	0.700	0.936	178	15
Studio2Shop with fc14 [20]	0.131	0.317	0.423	0.539	0.684	0.926	230	15
Studio2Shop with 128floats [20, 28]	0.132	0.319	0.426	0.540	0.677	0.909	274	15
Studio2Shop (with fDNA1.0) [20]	0.250	0.508	0.626	0.730	0.840	0.972	93	4
Studio2Shop [20] with fDNA 1.1	0.326	0.612	0.717	0.807	0.894	0.984	49	2
Fashion2Shop	**0.338**	**0.612**	**0.718**	**0.805**	**0.897**	**0.984**	**49**	**2**

Generally it was found in [20] that:

- Studio2Shop outperforms other architectures, mostly thanks to the non-linear matching module.
- fDNA1.0 can be replaced with any product representation such as fc14 (VGG16 features), but having a specialised feature representation makes a significant difference.
- The time needed for (naive) retrieval is too long for real-time applications, however it could be heavily shortened by pre-filtering the candidate products using a fast linear model and by optimising the implementation.

Note that FashionDNA 1.1 leads to a significant boost to the performance of Studio2Shop. In contrast, it seems that the segmentation and the image transformations have not made any further difference. It is unsurprising as the model cannot really perform better on our catalogue images simply because the background is white instead of neutral. Adding transformations could even make the task harder in principle, so it is rather reassuring that the performance is stable. It means that Street2Fashion works well enough on our studio images, despite the very few training instances it has seen. We will see the positive effect of the segmentation and image transformations on external datasets.

Retrieval. Figure 6 shows random examples of retrievals on test query images. The query image and its segmented version are shown on the left, while the top 50 suggestions are displayed on the right in western reading order. A few

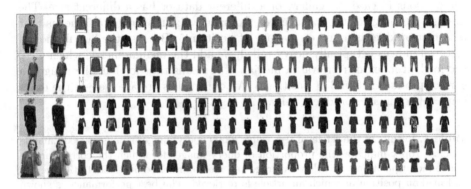

Fig. 6. Random examples of the retrieval test using 20000 queries against 50000 Zalando articles. Query images and their segmented version are in the left columns, next to two rows displaying the top 50 suggested products, in western reading order. Green boxes show exact hits. (Color figure online)

Table 5. Quantitative results on external datasets. Top-k retrieval indicates the proportion of query images for which a correct product was found at position lower than or equal to k. Average stands for the average retrieval position, median for the median retrieval position.

(a) DeepFashion In-Shop-Retrieval [25]. The dataset we put together to fit our setting contains 2922 query images associated with 683 products

	top-1	top-5	top-10	top-20	top-50	top-1%	average	median
Studio2Shop (with fDNA1.0) [20]	0.258	0.578	0.712	0.818	0.919	0.619	17	3
Studio2Shop [20] with fDNA 1.1	0.308	0.638	0.755	0.842	0.931	0.670	14	2
Street2Fashion2Shop	**0.336**	**0.675**	**0.785**	**0.867**	**0.941**	**0.703**	**12**	**2**

(b) LookBook [38]. The dataset we put together to fit our setting contains 68820 query images associated with 8726 products

	top-1	top-5	top-10	top-20	top-50	top-1%	average	median
Studio2Shop (with fDNA1.0) [20]	0.013	0.044	0.070	0.107	0.182	0.241	1266	466
Studio2Shop [20] with fDNA 1.1	0.018	0.054	0.083	0.125	0.202	0.263	1269	431
Street2Fashion2Shop	**0.031**	**0.094**	**0.141**	**0.199**	**0.298**	**0.374**	**759**	**197**

observations can be made. Firstly, we usually find one correct article very high up in the suggestions. Secondly, even if the correct article is not found, the top suggestions respect the style and are almost always relevant, which is the most important result as, realistically, the correct product will likely not be part of our assortment at query time. Secondly, we are able to retrieve more than one category. A more customer-friendly interface could exploit this to present the results in a more pleasing way, but if the image has a full-body, we usually find a mixture of tops and bottoms in the top suggested products.

6 Experiments on External Datasets

Most academic contributions whose data is publicly available focus on retrieving images which potentially contain models (and even backgrounds) on both sides, the query side and the product side. Practically speaking, a product can be represented by a variety of pictures, when in this manuscript the product is represented by FashionDNA based on title images only. Consequently, it is very difficult to compare our work to others, because most models are not reproducible, and most datasets do not fit our requirement, *i.e.* do not have isolated title images.

6.1 The Datasets

We use two external datasets.

– DeepFashion In-Sop-Retrieval [25]. The original study using DeepFashion [25] does not do any domain transfer, *i.e.* both query images and product images are of the same type, however the product images that are of the title kind can easily be isolated. We reduce the product set to those images, keeping only those from the DeepFashion categories that are closest to ours (Denim, Jackets_Vests, Pants, Shirts_Polos, Shorts, Sweaters, Sweatshirts_Hoodies, Tees_Tanks, Blouses_Shirts, Cardigans, Dresses, Graphic_Tees, Jackets_Coats, Skirts), leaving 683 products. The query set is then restricted to the images whose product is kept, *i.e.* 2922 images. We contacted the group to ask them to run their model on this reduced dataset so we could compare ourselves to them fairly, they did accept to help but never sent the results, even after several reminders on our part.
– LookBook [38]. LookBook is a dataset that was put together for the task of morphing a query image into the title image of the product it contains using generative adversarial networks. There are 68820 query images and 8726 products. The category of product is unknown so we keep all of them. Though the data is appealing because the product images are title images and the query images wild images, here again we cannot compare ourselves to them as they do not work on visual search.

6.2 Experimental Set-up

We apply FashionDNA 1.0 and 1.1 to the title images of these external product sets, without fine-tuning, to obtain the representations of the external products. For each dataset, we then apply Street2Fashion2Shop, again without finetuning, to the query images and match them against (a) the products coming from the dataset itself for quantitative assessment, (b) the Zalando products used in Sect. 5.4 for qualitative assessment.

6.3 Quantitative Results

Table 5 shows quantitative results for DeepFashion In-Shop-Retrieval (see Table 5a) and LookBook (see Table 5b). Top-k retrieval indicates the proportion of query images for which a correct product was found at position lower than or equal to k. Average stands for the average retrieval position, median for the median retrieval position. No fine-tuning is involved.

The segmentation gives a much greater boost to the performance on Look-Book images than on DeepFashion In-Shop-Retrieval images. This is to be expected as the DeepFashion images are clean and therefore do not benefit so much from the segmentation, while LookBook images are wild.

6.4 Qualitative Results

Figure 7 shows random examples of the qualitative experiments, for DeepFashion In-Shop-Retrieval (see Fig. 7a), LookBook (see Fig. 7b) and for street shots (see Fig. 7c). For each query image, the query image is displayed on the very left, followed by the segmented image and by the top 50 product suggestions. The reader may want to zoom into the figures to see the results better.

Without fine-tuning the model to this kind of data, the results are already pleasing. The functional category of the garments and the colour family are respected, which should be a minimum. The style usually follows too. For example, in LookBook's last example, the dress is not only blue but denim-like and so are most suggestions. In the first street shot, the man is wearing a sports outfit, and our model gives sports products as suggestions. Additionally, more than one categories are naturally returned when they are fully visible, which - if presented properly to customers - could be a nice feature. Note as well that in most cases (not all), although there is no specific garment detection, the model is not confused about which colour to assign to which garment, it has developed an internal intuition for where to find what, a prior on where garments usually locate so to say. An interesting example of this is the last street shot: our model does not know about scarves and there are no scarves in the assortment it could retrieve, but the scarf the man is wearing is located where the shirt should be, and our model suggests shirts of a similar pattern.

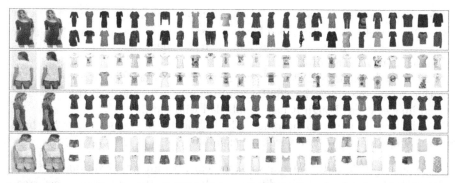

(a) Random examples of retrieval of Zalando products using query images from DeepFashion In-Shop-Retrieval [25].

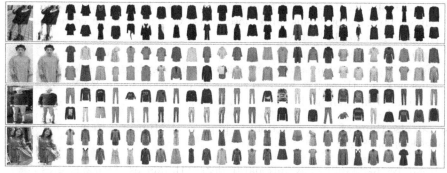

(b) Random examples of retrieval of Zalando products using query images from LookBook [38].

(c) Random examples of retrieval of Zalando products using query images from street shots.

Fig. 7. Qualitative results on external datasets. For each query image, the query image is displayed on the very left, followed by the segmented image and by the top 50 product suggestions. Better viewed with a zoom.

7 Conclusions and Future Work

We have presented Street2Fashion2Shop, a pipeline that enables us to train a visual search model using ill-suited but abundant annotated data, namely the studio images of our catalogue. The pipeline has three steps: a feature vector is obtained for all products using FashionDNA on their title image, the query image is segmented using Street2Fashion, then the query image and the products are matched using Fashion2Model.

Street2Fashion2Shop has the advantage to use a powerful product representation that most companies develop for multiple purposes, and to match products to wild images even though it has no labelled wild data.

Much work remains to be done. The segmentation model Street2Fashion could be improved in several ways, for example:

- It could be made fully convolutional to deal with images of various sizes.
- It is already able to deal with different scales thanks to the image transformations during training but it is not quite scale-invariant. We do not see much use in analysing shots taken form afar as it is not really a realistic use case, but we do have problems with close shots. Gathering more corresponding training data would be key.
- Each mask pixel is independent of the others in the loss. We could use a loss that enforces smoothness, for example by using regularisers on local regions. Alternatively, there has been work for example on using RNNs to model conditional random fields over the entire image [39].

We can also think of a few directions to improve the (image, product) matching model Fashion2Shop:

- It can be made fully convolutional to deal with images of various sizes.
- FashionDNA can be fine-tuned in the hope of needing only a simple linear match, instead of the current non-linear match which is much slower.
- The architecture of the query leg can be made more elegant and powerful, for example using ResNets as in FashionDNA 1.1.

All the small improvements mentioned above may bring performance gains, but they remain incremental efforts. A significant change we would like to make is not so much about performance but about flexibility. So far, a query image has to be matched against the whole assortment to find the most suitable products, which is not efficient. This can be made much faster by reducing the set of candidate products using a linear model to identify the most promising candidates and exploiting the structure of FashionDNA. Alternatively we would like to generate, from a query image, an appropriate distribution over FashionDNA. Sampling from this distribution would allow us to generate a set of relevant feature vectors that could then be matched to products using fast nearest neighbour search.

A   Image Transformations

Sections 4 and 5 mention applying image transformations during training to increase the robustness of the models. The parameters are chosen randomly within an appropriate range. These transformations are:

Fig. 8. Image transformations. The first row uses the original image while the second row uses the mirrored image. The columns represent the transformations and follow the order none, translate, rotate, zoom in, zoom out, blur, noise, darken and brighten.

- "none": no transformation applied.
- "translate": a translation "up or down" and "left or right". The (integer) vertical offset is sampled uniformly between -10% and 10% of the height of the image, the (integer) horizontal offset between -10% and 10% of the width, where the sign gives the direction of the shift.
- "rotate": a rotation. The (integer) angle is sampled uniformly between $[-10°, -3°] \cup [3°, 10°]$, where the sign gives the direction of the rotation.
- "zoom in": the image is magnified from its center. If the image is of size (h, w), a ratio r is uniformly sampled between 75% and 95%, then a rectangle of size $(r \times h, r \times w)$ is drawn around the center of the image, and the corners of this rectangle are stretched to the original corners of the image.
- "zoom out": the image is reduced from its center. If the image is of size (h, w), a ratio r is uniformly sampled between 75% and 95%, then a rectangle of size $(r \times h, r \times w)$ is drawn around the center of the image, and the original corners of the image are shrinked to the corners of this rectangle.
- "blur": the image is blurred using a Gaussian kernel of size uniformly sampled from the set $\{3, 5, 7\}$.
- "noise": Gaussian noise is added to the image with a variance sampled from a Gamma distribution with shape $= 1.1$ and scale $= 0.001/1.1$.
- "darken": the image is power-law transformed or γ-corrected with the inverse power γ uniformly sampled from the set $\{0.3, 0.4, 0.5, 0.6, 0.7, 0.8, 0.9\}$.
- "brighten": the image is power-law transformed or γ-corrected with the inverse power γ uniformly sampled from the set $\{1.3, 1.4, 1.5, 1.6, 1.7, 1.8, 1.9, 2, 2.1, 2.2, 2.3, 2.4, 2.5\}$.

Many of these methods are implemented using the OpenCV package for python. All these transformations can also be applied on the mirrored input image (obtained after a horizontal flip denoted "fliph" on the input image), such as fliph+darken for example, bringing a total of 18 transformations in our library.

Figure 8 shows examples of the alterations obtained using these transformations. If need be, any two or three of these transformations can be combined, for example "darken+rotate+blur", but this is out of scope. For each example, the first row uses the original image while the second row uses the mirrored image. The columns represent the transformations and follow the order none, translate, rotate, zoom in, zoom out, blur, noise, darken and brighten.

References

1. Cardoso, A., Daolio, F., Vargas, S.: Product characterisation towards personalisation: learning attributes from unstructured data to recommend fashion products. CoRR abs/1803.07679 (2018)
2. Bossard, L., Dantone, M., Leistner, C., Wengert, C., Quack, T., Van Gool, L.: Apparel classification with style. In: Lee, K.M., Matsushita, Y., Rehg, J.M., Hu, Z. (eds.) ACCV 2012, Part IV. LNCS, vol. 7727, pp. 321–335. Springer, Heidelberg (2013). https://doi.org/10.1007/978-3-642-37447-0_25
3. Bracher, C., Heinz, S., Vollgraf, R.: Fashion DNA: merging content and sales data for recommendation and article mapping. CoRR abs/1609.02489 (2016)
4. Chen, H., Gallagher, A., Girod, B.: Describing clothing by semantic attributes. In: Fitzgibbon, A., Lazebnik, S., Perona, P., Sato, Y., Schmid, C. (eds.) ECCV 2012, Part III. LNCS, vol. 7574, pp. 609–623. Springer, Heidelberg (2012). https://doi.org/10.1007/978-3-642-33712-3_44
5. Chen, Q., Huang, J., Feris, R., Brown, L.M., Dong, J., Yan, S.: Deep domain adaptation for describing people based on fine-grained clothing attributes. In: The IEEE Conference on Computer Vision and Pattern Recognition (CVPR) (2015)
6. Di, W., Wah, C., Bhardwaj, A., Piramuthu, R., Sundaresan, N.: Style finder: fine-grained clothing style detection and retrieval. In: The IEEE Conference on Computer Vision and Pattern Recognition (CVPR) Workshops (2013)
7. Dong, J., Chen, Q., Xia, W., Huang, Z., Yan, S.: A deformable mixture parsing model with parselets. In: ICCV, pp. 3408–3415 (2013)
8. Fu, J., Wang, J., Li, Z., Xu, M., Lu, H.: Efficient clothing retrieval with semantic-preserving visual phrases. In: Lee, K.M., Matsushita, Y., Rehg, J.M., Hu, Z. (eds.) ACCV 2012, Part II. LNCS, vol. 7725, pp. 420–431. Springer, Heidelberg (2013). https://doi.org/10.1007/978-3-642-37444-9_33
9. He, K., Gkioxari, G., Dollár, P., Girshick, R.: Mask R-CNN. In: Proceedings of the International Conference on Computer Vision (ICCV) (2017)
10. He, K., Zhang, X., Ren, S., Sun, J.: Deep residual learning for image recognition. CoRR abs/1512.03385 (2015)
11. Heinz, S., Bracher, C., Vollgraf, R.: An LSTM-based dynamic customer model for fashion recommendation. In: Proceedings of the 1st Workshop on Temporal Reasoning in Recommender Systems (RecSys 2017), pp. 45–49 (2017)
12. Huang, J., Feris, R.S., Chen, Q., Yan, S.: Cross-domain image retrieval with a dual attribute-aware ranking network. In: IEEE International Conference on Computer Vision, ICCV 2015, Santiago, Chile, 7–13 December 2015, pp. 1062–1070 (2015)
13. Jagadeesh, V., Piramuthu, R., Bhardwaj, A., Di, W., Sundaresan, N.: Large scale visual recommendations from street fashion images. In: Proceedings of the 20th ACM SIGKDD International Conference on Knowledge Discovery and Data Mining, KDD 2014, pp. 1925–1934 (2014)
14. Jetchev, N., Bergmann, U.: The conditional analogy gan: swapping fashion articles on people images. In: The IEEE International Conference on Computer Vision (ICCV) Workshops, October 2017

15. Ji, X., Wang, W., Zhang, M., Yang, Y.: Cross-domain image retrieval with attention modeling. In: Proceedings of the 2017 ACM on Multimedia Conference, MM 2017, pp. 1654–1662 (2017)
16. Jing, Y., et al.: Visual search at pinterest. In: KDD, pp. 1889–1898 (2015)
17. Kalantidis, Y., Kennedy, L., Li, L.J.: Getting the look: clothing recognition and segmentation for automatic product suggestions in everyday photos. In: Proceedings of the 3rd ACM Conference on International Conference on Multimedia Retrieval, ICMR 2013, pp. 105–112 (2013)
18. Kiapour, M.H., Yamaguchi, K., Berg, A.C., Berg, T.L.: Hipster wars: discovering elements of fashion styles. In: Fleet, D., Pajdla, T., Schiele, B., Tuytelaars, T. (eds.) ECCV 2014, Part I. LNCS, vol. 8689, pp. 472–488. Springer, Cham (2014). https://doi.org/10.1007/978-3-319-10590-1_31
19. Krizhevsky, A., Sutskever, I., Hinton, G.E.: ImageNet classification with deep convolutional neural networks. In: Advances in Neural Information Processing Systems 25, pp. 1097–1105 (2012)
20. Lasserre, J., Rasch, K., Vollgraf, R.: Studio2shop: from studio photo shoots to fashion articles. In: International Conference on Pattern Recognition, Applications and Methods (ICPRAM) (2018)
21. Liang, X., et al.: Deep human parsing with active template regression. IEEE Trans. Pattern Anal. Mach. Intell. **37**, 2402–2414 (2015)
22. Liang, X., et al.: Human parsing with contextualized convolutional neural network. In: ICCV, pp. 1386–1394 (2015)
23. Liu, S., et al.: Hi, magic closet, tell me what to wear! In: Proceedings of the 20th ACM International Conference on Multimedia, MM 2012, pp. 619–628 (2012)
24. Liu, S., Song, Z., Liu, G., Xu, C., Lu, H., Yan, S.: Street-to-shop: cross-scenario clothing retrieval via parts alignment and auxiliary set. In: IEEE Conference on Computer Vision and Pattern Recognition (CVPR), pp. 3330–3337 (2012)
25. Liu, Z., Luo, P., Qiu, S., Wang, X., Tang, X.: DeepFashion: powering robust clothes recognition and retrieval with rich annotations. In: Proceedings of IEEE Conference on Computer Vision and Pattern Recognition (CVPR) (2016)
26. Ronneberger, O., Fischer, P., Brox, T.: U-Net: convolutional networks for biomedical image segmentation. In: Navab, N., Hornegger, J., Wells, W.M., Frangi, A.F. (eds.) MICCAI 2015. LNCS, vol. 9351, pp. 234–241. Springer, Cham (2015). https://doi.org/10.1007/978-3-319-24574-4_28
27. Shankar, D., Narumanchi, S., Ananya, H.A., Kompalli, P., Chaudhury, K.: Deep learning based large scale visual recommendation and search for e-commerce. CoRR abs/1703.02344 (2017)
28. Simo-Serra, E., Ishikawa, H.: Fashion style in 128 floats: joint ranking and classification using weak data for feature extraction. In: Proceedings of the Conference on Computer Vision and Pattern Recognition (CVPR) (2016)
29. The Guardian (Zoe Wood): Asos app allows shoppers to snap up fashion (2017), https://www.theguardian.com/business/2017/jul/15/asos-app-allows-shoppers-to-snap-up-fashion
30. Vittayakorn, S., Yamaguchi, K., Berg, A.C., Berg, T.L.: Runway to realway: visual analysis of fashion. In: IEEE Winter Conference on Applications of Computer Vision, pp. 951–958 (2015)
31. Wang, N., Haizhou, A.: Who blocks who: simultaneous clothing segmentation for grouping images. In: Proceedings of the International Conference on Computer Vision, ICCV 2011 (2011)

32. Wang, X., Sun, Z., Zhang, W., Zhou, Y., Jiang, Y.G.: Matching user photos to online products with robust deep features. In: Proceedings of the 2016 ACM on International Conference on Multimedia Retrieval, ICMR 2016, pp. 7–14 (2016)
33. Wang, X., Zhang, T.: Clothes search in consumer photos via color matching and attribute learning. In: Proceedings of the 19th ACM International Conference on Multimedia, MM 2011, pp. 1353–1356 (2011)
34. Yamaguchi, K., Kiapour, M.H., Berg, T.L.: Paper doll parsing: retrieving similar styles to parse clothing items. In: IEEE International Conference on Computer Vision, pp. 3519–3526 (2013)
35. Yamaguchi, K., Kiapour, M.H., Ortiz, L., Berg, T.: Parsing clothing in fashion photographs. In: Proceedings of the 2012 IEEE Conference on Computer Vision and Pattern Recognition, CVPR 2012, pp. 3570–3577 (2012)
36. Yamaguchi, K., Okatani, T., Sudo, K., Murasaki, K., Taniguchi, Y.: Mix and match: joint model for clothing and attribute recognition. In: Proceedings of the British Machine Vision Conference (BMVC), pp. 51.1–51.12 (2015)
37. Yang, F., et al.: Visual search at ebay. In: Proceedings of the 23rd ACM SIGKDD International Conference on Knowledge Discovery and Data Mining, KDD 2017, pp. 2101–2110 (2017)
38. Yoo, D., Kim, N., Park, S., Paek, A.S., Kweon, I.S.: Pixel-level domain transfer. In: Leibe, B., Matas, J., Sebe, N., Welling, M. (eds.) ECCV 2016. LNCS, vol. 9912, pp. 517–532. Springer, Cham (2016). https://doi.org/10.1007/978-3-319-46484-8_31
39. Zheng, S., et al.: Conditional random fields as recurrent neural networks. In: Proceedings of the 2015 IEEE International Conference on Computer Vision (ICCV), ICCV 2015, pp. 1529–1537 (2015)
40. Zhu, S., Fidler, S., Urtasun, R., Lin, D., Loy, C.C.: Be your own prada: fashion synthesis with structural coherence. In: International Conference on Computer Vision (ICCV) (2017)

Interactive Design Support for Architecture Projects During Early Phases Based on Recurrent Neural Networks

Johannes Bayer[✉], Syed Saqib Bukhari, and Andreas Dengel

German Research Center for Artificial Intelligence,
Trippstadter Strasse 122, 67663 Kaiserslautern, Germany
{johannes.bayer,saqib.bukhari,andreas.dengel}@dfki.de
https://www.dfki.de

Abstract. In the beginning of an architectural project, abstract design decisions have to be made according to the purpose of the later building. Based on these decisions, a rough floor plan layout is drafted (and subsequently redrafted in successively more refined versions). This entire process can be considered an iterative design algorithm, in which high-level ideas and requirements are transformed into a specific building description.

Nowadays, this process is usually carried out in a manual and labor-intensive manner. More precisely, concepts are usually drafted on semi-transparent paper with pencils so that a when a new sheet of paper is put on an existing one, the old concept may serve as a template for the next step in the design iteration.

In this paper, we present a semi-automatic approach to assist the developer by proposing suggestions for solving individual design steps automatically. These suggested designs can be modified between two successive automatic design steps, hence the developer remains in control of the overall design process. In the presented approach, floor plans are represented by graph structures and the developer's behavior is modeled as a sequence of graph modifications. Based on these sequences we trained a recurrent neural network-based predictor that is used to generate the design suggestions. We assess the performance of our system in order to show its general applicability.

The paper at hand is a extended version of our ICPRAM 2018 conference paper [1], in which we address the different aspects of our proposed algorithm, challenges we faced during our research as well as intended work flow in greater detail.

Keywords: Interactive design support · Early phase support
Architecture project · LSTM · Archistant

© Springer Nature Switzerland AG 2019
M. De Marsico et al. (Eds.): ICPRAM 2018, LNCS 11351, pp. 27–43, 2019.
https://doi.org/10.1007/978-3-030-05499-1_2

1   Introduction

In many architectural projects, the development of a building can be considered to be following a top-down strategy. In the entire life-cycle of a building (formulation of an initial concept of a building to its demolition), the early design phase deals with deriving a first and rough floor plan layout given an abstract idea. Such a concept may be purpose-driven (e.g. 'apartment building for 12 average families with children in a mid-sized city') or may contain stylistic wishes ('Tuscan Mansion') as well as semiotic statements ('The temple should reflect the openness of our religion'). Given that such a high-level description usually comes from a customer or contractor, this initial concept can also be considered a requirement to the project and therefore contains rather punctual yet specific constrains ('The house should have two bathrooms as well as three sleeping rooms and the living room should be next to the kitchen'). Generally, only a rough idea is given, hence most of the final floor plan's aspects remain vague.

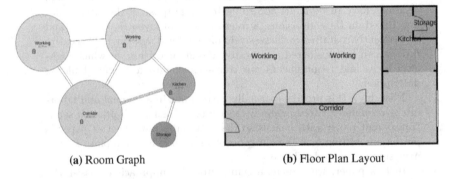

(a) Room Graph   (b) Floor Plan Layout

Fig. 1. Illustration of the room schedule work flow (a → b).

Buildings can be described using graphs structures. In such a modeling, individual rooms are conveniently described as nodes while the connections between them (e.g. walls, doors) are described as edges. Using such a description, all properties of a floor plan layout must consequently be expressed as properties of nodes and properties of edges. The surrounding walls of a room for example can be denoted as a polygon that is a property of the node which represents the room.

Given a graph-based description of a building, the early-phase design task in architecture translates into an iterative graph manipulation algorithm (see Fig. 1). This technique is already existing as a traditional and manual working method in architecture, the so-called *room schedule* (also referred to as *architectural program*). From the algorithmic viewpoint, the building project's concept (or requirements) translate to structural constrains of the building's graph at the start of the algorithm: A fixed amount of rooms translates to a fixed amount of nodes in the graph. Adjacency requirements ('The kitchen must be next to the living room') translate to predetermined edges in the graph. Defined room aspects translate to fixed properties of the corresponding nodes. The task of such an algorithm is to complete the missing graph properties.

The entire process described to far is nowadays usually carried out manually and informally. This usually involves semi-transparent sketching paper that is being written on with pencils. Algorithmically, each sheet of semitransparent paper represents one iteration (or corresponds to a set of graph modifications). When putting on a new sheet of semi-transparent paper on top of the old one, the old sketch serves as a template for the next, refining design iteration. While a fair amount of this task is highly creative, many actions remain rather repetitive and monotone for the architectural developer. Usually, this process makes use of previously existing projects since architectural projects often work with references. Hence, graph structures from former projects are assumed to be at least partly copied or resembled in newer ones.

In this paper, a semi-automated approach for drafting architectural sketches is described. In Sect. 2, the existing base technologies and concepts on which the approach of this paper is based are outlined. In Sect. 3, the employed representation of floor plans for recurrent neural networks is introduced. Section 4 describes the central machine learning approach, i.e. how models are trained and how floor plan drafts are extended as well as completed using the trained models. Some of the problems encountered during the design of the system as well as how trade-offs attempt to address them are also described here. After that, in Sect. 5 the integration of the approach into an existing sketching software is outlined. The approach is evaluated in Sect. 6, where the results of an automatically conducted performance evaluation on a set of floor plans is presented. Examples of real-value outputs of a trained models are provided to illustrate the use of the integrated system. Section 7 concludes this paper by giving an outlook on possible future research directions.

2 Related Work

2.1 The Long-Short Term Memory

Long-Short Term Memories [7,9] are a class of recurrent artificial neural networks. During each time step, they are supplied with an input vector of arbitrary (but fixed) length, while they emit an output vector of a size equivalent to their amount of cells. Stacked with an MLP, they may also return a vector of also arbitrary (but fixed) size. As vector sequence processing units, LSTMs possess different key properties like the ability to transform data in various ways and storing information for an arbitrary length of time steps. The components of their input vectors have to be normalized to a certain the interval (often $[0,1]$). Likewise, their output values are limited to a certain interval (often $[0,1]$). The experiments described in this paper have been conducted using the OCRopus LSTM implementation [4].

2.2   The Architectural Design Support Tool Archistant

Archistant [11] is an experimental system for supporting architectural developers
during early design phases. Its key feature is the search of floor plans similar to an
entered sketch. It consists of a front-end, the Archistant WebUI, and a modular
back-end, in which floor plans are processed between the individual modules via
the dedicated AGraphML format.

Archistant WebUI. The Archistant WebUI (see Fig. 2, formerly known as
Metis WebUI [2]) is the graphical user interface of the system. It mainly consists
of a sketch editor for floor plans (originally the created sketches just served as
search requests to the core system). The workflow intended by the Archistant
WebUI follows the traditional room schedule working method. Every aspect of
a room may be specified as abstract or specific as intended by the user and
the degree of abstractness may be altered by the user during his work. This

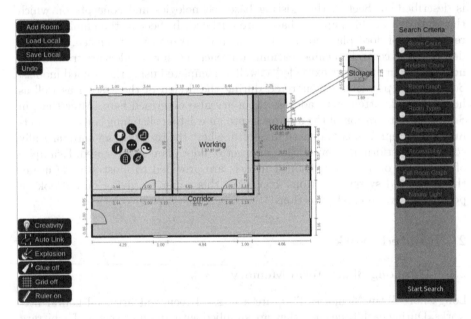

Fig. 2. Screenshot of the Archistant WebUI.

Table 1. Edge types in Archistant.

Type	Description	Visualization (WebUI)
Wall	Rooms share a uninterrupted wall only	1 Continuous Line
Door	Rooms connected by door	2 Continuous Lines
Entrance	Rooms connected via a reinforced door	2 Dashed Lines
Passage	Rooms connected by a simple discontinuity in a wall	3 Continuous Lines

continuous refinement allows for a top-down work process, in which a high-level building description is transformed into a specific floor plan. The Archistant WebUI comes as a web application and therefore runs inside a HTML5-supporting web browser.

AGraphML. AGraphML [10] is Archistant's exchange format for floor plan concepts. AGraphML itself is a specification of GraphML [3]. It follows the convention of the paper at hand, in which rooms are modeled as nodes and their connections are modeled as edges. The AGraphML specification therefore mainly consists of the definitions of node and edge attributes (see Table 1).

3   Encoding Floor Plans for Recurrent Neural Network Processing

This section outlines the representation of floor plans which is used to make them processable by recurrent neural networks. In the current status of our work, we restrict ourselves to a limited set of floor plan attributes that are incorporated into this representation: Room functions, connections between rooms, room layouts (i.e. a polygon which is representing a rooms surrounding walls), and information whether or not natural light is available in a room or not (which roughly equals to whether or not a room is equipped with at least one window).

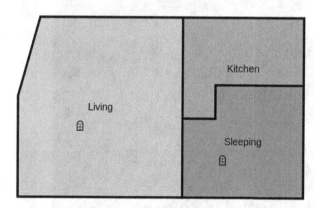

Fig. 3. Rendered image of a sample floor (from [1]). Window symbols indicate access to natural light. The detail level shown in this image equals the information contained in the neural network representation.

3.1   Requirements

In order to be processable by recurrent neural networks, floor plans need to be described as a sequence of feature vectors (all feature vectors must have the same length). There are several requirements to this sequence representation: Both sequence length and vector size should be as small as possible in order to minimize learning and inference execution times. The vectors should be easy to interpret by automated means. Finally, the actual information should be organized into separated chunks of small vector sequences and these should be separated by data-less so-called *control vectors* (their purpose will be explained later). Most important, the information flow in the vector sequence should mimic the actual workflow of a user who develops a floor plan. Consequently, abstract information should precede specific information, i.e. declaration of all rooms along with their room functions should be before the declaration of the actual room layouts.

3.2 Blocks

A complete floor plan description in the chosen representation consists of 3 consecutive blocks:

1. Room Function Declarations.
2. Room Connections.
3. Room Geometry Layouts.

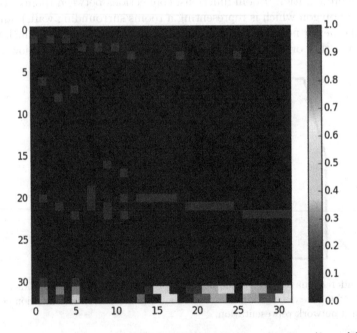

Fig. 4. Feature vector sequence of the same sample floor encoding (from [1]). Every feature vector occupies one column. The encoding consists of three blocks. The first block ranges from column 0 to 5 and defines the rooms along with IDs (row 20–29), the room function, the room's center position (row 31 and 32), and whether or not a room has access to natural light (row 30). In Block 2 (column 6–12) connections between the rooms are defined. In block 3 (column 13–31) the polygons of the room surrounding walls are described.

Each block consists of a number of tags of the same kind (i.e. there is one tag type for each block). Each tag is represented by a number of vectors. An example for a rendered floor plan along with its representation as a sequence of vectors can be seen in Figs. 3 and 4 respectively.

3.3   The Feature Vector

The feature vector is considered to be structured into several channels (see Fig. 5 for the relation between channels and actual feature vector components):

– The blank channel indicates that no information are present (used to indicate start and ending of floor plans or to signal the LSTM to become active)
– The control channel indicates that a new tag begins and what type the new tag is
– The room type channel (room types are called room functions in architecture)
– The connection type channel
– The room ID channel is used to declare or reference an individual room
– A second ID channel is used for connection declaration
– has Window property

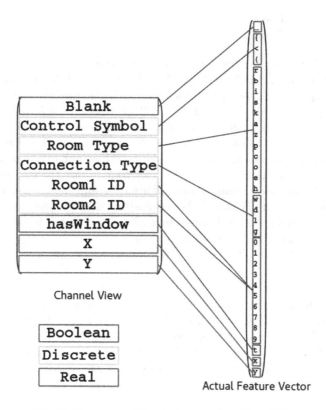

Fig. 5. Structure of the feature vector (from [1]).

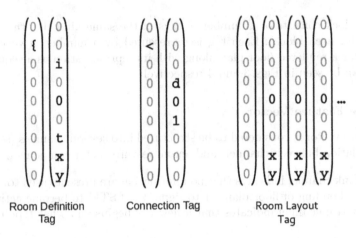

Room Definition Tag   Connection Tag   Room Layout Tag

Fig. 6. The different tag types in feature vectors representation (channel view, from [1]). Left: Definition of a Living (i) Room with ID (1) with a Window (t) with a center at position (x, y). Middle: Definition of a door connection (d) between rooms 0 and 1. Right: Definition of the polygon layout of room 0.

- X Ordinate of a Point
- Y Ordinate of a Point

Generally, LSTMs allow for real-valued vector components as input and output. However, only a 2D-Point's X and Y ordinates (which are both normalized to $[0, 1]$) actually make use of this capability. All other components are modeled as boolean variables (0.0 for `false`, 1.0 for `true`). In a former version of the floor plan representation, 1-Hot Encoding has been considered. However, this approach is rather inefficient compared to the finally chosen one (the final, tighter encoding occupies around 88 less space than our experimental 1-Hot encoding of the same information). So in order to minimize the sequence length, each feature vector may contain several information and the ID of each room is represented as a dedicated channel. As an disadvantage, this finally chosen, tight encoding only allows for a fixed upper limit of rooms that has to be determined before training (10 rooms in the paper at hand).

3.4 Tags

Tags are the atomic units of floor plan features. At the same time, they can be considered graph modifications carried out by an entity (like the user) to develop a floor plan. Every tag is represented by a set of successive feature vectors. All tags start with a so-called control vector, that only serves the purpose of indicating the tag's type. Figure 6 illustrates how the different tag types are made up from feature vectors. Currently, there are three different types of tags:

Room Definition Tags. Each tag of this type defines a room by assigning it an ID along with a room type, a flag indicating whether or not the room has a window, and the position of the room's center. This tag type always occupies 2 feature vectors.

Connection Tags. A Connection Tag declares a connection between two rooms. It consists of the references between the two connection partners as well as the connection type. Since all room IDs are represented in the feature vector by dedicated components, two room IDs can be represented in one feature vector. The supported connection types are taken from the definition of Archistant. This tag type always occupies 2 feature vectors.

Room Layout Tags. These tags define a polygon surrounding walls around a room by describing the individual corners of the room's walls successively. A room layout tag occupies $p + 1$ feature vectors, where p is the number of corners of the room.

3.5   Room and Connection Order

The order of rooms and connections in the floor plan representation underlies a trade-off: a well defined order of rooms and connections only allows for exactly one representation of the same floor plan. By allowing for any arbitrary order of rooms and connections, the actual user behavior is better approximated and there are multiple representations of the same floor plan (many samples can be created from one single floor plan). However, when considering elements to be given in a random fashion, an LSTM that should predict them is difficult to train. Since a random order of room definitions and connections adds an unpredictable noise to the LSTM, the order or rooms and connections is defined as follows:

Room Order. The order in which the rooms are given in the feature vector sequence is determined by the center position of the room within the floor plan. A room which center has a smaller X ordinate appears before a room with a greater X center ordinate. In case of the centers of two rooms share the same X ordinate, the room with the smaller Y ordinate precedes the other room (top to bottom, left to right). The order of rooms is the same for block 1 and block 3.

Connection Order. The order of connections in the feature vector sequence is determined by the order of the rooms. At this point, the connection graph is considered to be directed and that the source room IDs are always smaller than target room IDs. If two connections have different source room IDs, the connection with the lower source room ID will come before the one with the higher source room ID. If two connections have the same source room ID, the connection with the lower target room ID will precede the connection with higher target room ID.

4   Proposed Mechanism of Autocompletion of Floor Plans Using LSTM

In this section, the proposed algorithm for expanding and completing floor plans is presented. More precisely, the modus operandi, in which existing parts of floor plans are given to the (LSTM) model and new floor plan parts are retrieved is outlined.

4.1   Input and Output Sequences

The structure of the model's input vector is identical to the structure of its output vector and therefore both are referred to as feature vectors. Consequently and as a general working principle, existing parts of a floor plan are used as input to the model while new floor plan parts are retrieved from the model's output. Two different approaches that implement such a behavior are examined here: *block generation sequencers* and *vector prediction sequencers*.

Block Generation Sequencers. Block generation sequencers follow a simple pattern: The first n blocks are given to the model's input. Simultaneously, the model's output is simply a series of blank vectors (the blank component is 1.0, while all other components remain 0.0). Afterwards, a sequence of blank vectors is used as input while the $n + 1th$ block is emitted by the model (eventually finished by a blank vector). As a result, there has to be one model trained for each block that should be predicted. Consequently, multiple models are used to support the user during the entire work flow. Additionally, a model for supporting the first design step cannot be created using this kind of sequencer.

Vector Prediction Sequencers. Vector prediction sequencers aim to predict the n-th vector of a sequence given the first $n - 1$ sequence vectors. These sequencers are trained by using a concatenation of a blank vector with the full feature vector sequence of a floor plan as the input. As an output, the full feature vector sequence of the same floor plan is used, but this time followed by a blank vector.

4.2   Preparation of Database

In order to make maximum use of the limited set of floor plans available in AGraphML, some preprocessing is applied to generate the final training set as well as test set (see Fig. 7). The validation set is omitted here since the amount of floor plans available was very limited and previous experiments on a similar DB indicated that overfitting is not a serious issue in the given situation. First of all, the original sample set is split into two disjoint subsets. Each floor plan is now converted into b different samples (we refer to this process as blow-up and to b as blow-up factor). A sample is derived from a floor plan by rotating all point

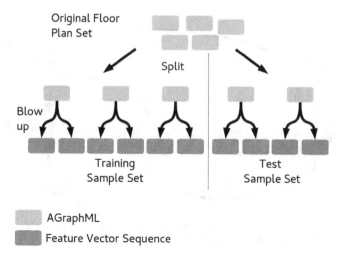

Fig. 7. Sample preparation (from [1]).

of the floor plan (centers, corner points) by an random vector and then create the feature vector sequence as described so far. During this step, the center and corner points of the floor plan are also normalized to the $[0, 1]^2$ space. Because of the applied rotation, the order of rooms and connections differs within the samples generated from one single floor plan.

4.3 Extension of Floor Plans

The two different sequencing approaches need different strategies to generate new floor plan aspects, as outlined below:

Block Generation Sequencers. In this approach, a new block is generated by feeding a concatenation of previous blocks with a sequence of blank vectors into the model and reading the predicted block from the model's output. The blank vector sequence length must be larger that the expected length of the predicted block. This is can be done by determining the upper limit of predicted block length in the training database. A more sophisticated approach is to make use knowledge on the input sketch (e.g. the room count of the input sketch).

Vector Prediction Sequencers. Following a metaphor by Alex Graves, in which sequence predictors used for sequence generation are compared to a person dreaming (both are iteratively treating their own output as new input [8]), this structures works like a dreaming person who occasionally gets inspiration from outside and who combines the information from outside with its flow of dreaming. Because of that, this technique is referred to as the *shallowDream* structure here (see Fig. 8).

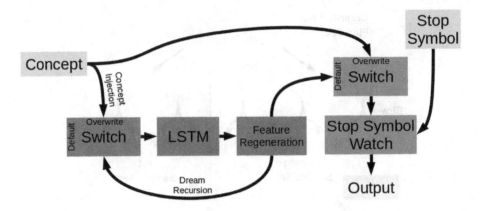

Fig. 8. The shallowDream structure (from [1]). The inputs are marked green. Components of the LSTM recursion are marked blue. (Color figure online)

Basically, this structure operates in two different phases. During the first phase, the already existing floor plan parts as provided or accepted by the user (also referred to as *concept*) are fed into the model. During this phase of *concept injection*, all outputs of the model are ignored (the concept is kept as it is in the feature vector sequence, i.e. existing requirements or design iterations are fully sustained). After the concept has been injected completely, the model takes over both the generation of the structure's output and also serves as its own input. This phase is also referred to as *generation phase*. The process is terminated when a predefined stop symbol occurs in the feature vector sequence.

Using the shallowDream structure, it is possible to implement multiple different functions by simply altering the concept and the stop symbol. For example, in order to predict room connections, the a concatenation of block 1 and the control vector of a block 2 tag (connection tag) is used as concept and the control vector of a room layout tag is used as stop symbol. The control vector at the end of a concept is used to instruct the model to generate the favored tag type (and hence to start the new block).

Even after intensive training, the output produced by the model only approximates the intended feature vectors. An error at any time step influences the outputs of all subsequent time steps due to the recursive characteristics of the shallowDream structure. This effect aggravates over time, therefore feature vectors have to be regenerated during the generation phase. For that purpose, three different strategies are proposed:

No Regeneration. In this primitive approach, the current feature vector is simply reinserted without any modifications into the models input.

Vector-Based Regeneration. This strategy solely utilizes knowledge about the feature vector's structure. Generally, all boolean components are recovered by mapping them to 1.0 or 0.0 based on which the component is closer to and the real-valued components remain unaltered.

Sequence-Based Regeneration. In this approach, a state machine is keeping track of the current block and tag the sequence is in (thereby utilizing knowledge about the sequence structure). Based on that information a vector is regenerated by calculating the most likely, possible vector. In order for the state machine to transit between different states, only selected components are evaluated against a threshold. Different components (and combinations of them) might be used in different blocks.

5   Integration of the Proposed Mechanism into Archistant

This paper is restricted to the following two functions:

- Room Connection Generation. Given a set of rooms (each room is described by a center position coordinate and room function) connections are generated between them, turning a set of rooms into a room graph.
- Room Layout Generation. Given a room Graph, layouts for each room a layout (i.e. a polygon describing its surrounding walls) is generated.

For the sake of simplicity, a single button is added to the WebUI only, which we labeled "Creativity". Based on the current state of the user's work, the different functions are selected automatically.

6   Experiments

LSTMs are trained based on the two described sequencing approaches. In all cases, a training database with 200 entries, a test database of 40 entries, a blowup factor of 30, 500 LSTM cells and a learning rate of 0.01 are used.

6.1   Quantitative Analysis

In order to compare the performances of the different approaches, the room connection generation is calculated on the room definitions of the test set of floor plans and compared with these floor plan's actual connections. As a metric, the amount of wrong connections is divided by the amount of actual connections. A connection is considered to be completely wrong (error 1.0), if the predicted connection does not actually exist in the ground truth. It is considered partly wrong (error 0.5), if the connection type in ground truth is different. The final evaluation value is the arithmetic average over all floor plans in the test set. The results are shown in Table 2. It is emphasized that floor plan generation is a creative task and that there is not necessarily one solution to a given problem, i.e. the used error calculation metrics do only hint the actual performance of the approaches (an error of 0% appears unrealistic to accomplish given that also the floor plans in the database are only one way to solve the problem).

Table 2. Performance comparison of the individual approaches for the connection generation task on the test set (from [1]).

Approach	Error
Block generation sequencer	65.78%
Vector prediction sequencer	66.08%

6.2   Qualitative Analysis

The performance of the shallowDream structure is shown exemplary for two scenarios. For the room connection generation, four rooms are given (Fig. 10) as a concept. As illustrated in Fig. 9, the regeneration strategy influences the output. Figure 11 depicts a rendered version of the output produced by the sequence based regeneration.

In order to illustrate the performance of the room layout generation function, a different starting situation is used (see Fig. 12), the result is depicted in Fig. 13.

7   Future Work

We have shown the general viability of our approach (i.e. the output of the trained models resembles the intended syntax in a quality sufficient for our inferencing algorithm to produce results). Nevertheless, a lot of floor plan aspects are not yet covered in the existing models. The actual position of doors and windows are needs to be included into the models as well as support for multi-storey buildings. Apart from that, the performance of the existing models is still fairly limited. At the moment, there is only one phase of concept injection followed by a generation phase. By allowing for multiple alternating phases of generation and concept injection, even more functions could be realized. Apart from that, better metrics have to be found to assess a models performance. As of now, our approach only supports a strict order of design phases. A more flexible approach that allows for a more fluid transition between design steps would be useful in future.

In order to both allow for better comparison to similar approaches and to improve the performance of our system, the presented approach can be applied to a standard database [5] and existing algorithms [6] can be used to increase the sample size of the sample database.

Many of the mentioned problems could be avoided by a completely different sequencing strategy: Instead of employing a single neural network that manipulates the entire room graph, multiple neural network instances could be used instead. In such a scenario, one LSTM could be responsible for a single room. Such a group of LSTMs could communicate the results of individual design steps among each other. Not only would this approach allow for shorter sequences, it would also render most of the room ID assignment as well as room and connection ordering redundant. The problem of a predetermined, fixed upper limit of

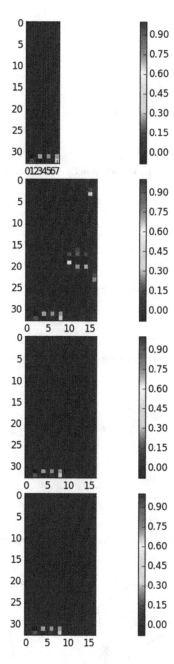

Fig. 9. The regeneration technique in shallowDream influences the generated feature vector sequences (from [1]). From Top to Bottom: 1. The concept. 2. No Regeneration. 3. Vector-Based Regeneration 4. Sequence-Based Regeneration.

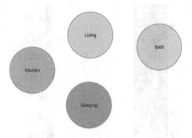

Fig. 10. Input given to the shallowDream structure (from [1]).

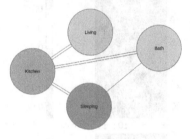

Fig. 11. Output obtained from the shallowDream structure (from [1]).

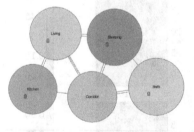

Fig. 12. Input given to the shallowDream structure (from [1]).

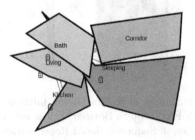

Fig. 13. Output obtained from the shallowDream structure (from [1]).

rooms would therefore also be mitigated tremendously in such a setting. Likewise, one of the assumed reasons for the low training performance of the models is the amount of information that the LSTMs had to store. This problem would also be mitigated in a distributed approach.

Apart from that, our approach can be used as a general template for machine learning of user behavior, given that the data structure manipulated by the user can be described as a graph. Consequently, a more general implementation of a graph-based machine learning framework can be build using our approach.

Acknowledgment. This work was partly funded by Deutsche Forschungs-Gemeinschaft.

References

1. Bayer, J., Bukhari, S., Dengel, A.: Interactive LSTM-based design support in a sketching tool for the architectural domain. In: 7th International Conference on Pattern Recognition Applications and Methods, Funchal (2018)
2. Bayer, J., et al.: Migrating the classical pen-and-paper based conceptual sketching of architecture plans towards computer tools - prototype design and evaluation. In: Lamiroy, B., Dueire Lins, R. (eds.) GREC 2015. LNCS, vol. 9657, pp. 47–59. Springer, Cham (2017). https://doi.org/10.1007/978-3-319-52159-6_4
3. Brandes, U., Eiglsperger, M., Lerner, J., Pich, C.: Graph Markup Language (GraphML). In: Tamassia, R. (ed.) Handbook of Graph Drawing and Visualization, vol. 20007, pp. 517–541. CRC Press, Boca Raton (2013)
4. Breuel, T.M.: The OCRopus open source OCR system. In: Electronic Imaging 2008, p. 68150F. International Society for Optics and Photonics (2008)
5. de las Heras, L.-P., Terrades, O.R., Robles, S., Sánchez, G.: CVC-FP and SGT: a new database for structural floor plan analysis and its groundtruthing tool. Int. J. Doc. Anal. Recognit. (IJDAR) **18**(1), 15–30 (2015)
6. Delalandre, M., Pridmore, T., Valveny, E., Locteau, H., Trupin, E.: Building synthetic graphical documents for performance evaluation. In: Liu, W., Lladós, J., Ogier, J.-M. (eds.) GREC 2007. LNCS, vol. 5046, pp. 288–298. Springer, Heidelberg (2008). https://doi.org/10.1007/978-3-540-88188-9_27
7. Gers, F.A., Schmidhuber, J., Cummins, F.: Learning to forget: continual prediction with LSTM. Neural Comput. **12**(10), 2451–2471 (2000)
8. Graves, A.: Generating sequences with recurrent neural networks. arXiv preprint arXiv:1308.0850 (2013)
9. Hochreiter, S., Schmidhuber, J.: Long short-term memory. Neural Comput. **9**(8), 1735–1780 (1997)
10. Langenhan, C.: Datenmanagement in der Architektur. Dissertation, Technische Universität München, Müchen (2017)
11. Sabri, Q.U., Bayer, J., Ayzenshtadt, V., Bukhari, S.S., Althoff, K.-D., Dengel, A.: Semantic pattern-based retrieval of architectural floor plans with case-based and graph-based searching techniques and their evaluation and visualization (2017)

CNN-Based Deep Spatial Pyramid Match Kernel for Classification of Varying Size Images

Shikha Gupta[1]([✉]), Manjush Mangal[1], Akshay Mathew[1],
Dileep Aroor Dinesh[1]([✉]), Arnav Bhavsar[1], and Veena Thenkanidiyoor[2]

[1] School of Computing and EE, Indian Institute of Technology, Mandi, H.P., India
{shikha_g,manjush_mangal,akshay_mathew}@students.iitmandi.ac.in
{addileep,arnav}@iitmandi.ac.in
[2] Department of CSE, National Institute of Technology Goa, Ponda, India
veenat@nitgoa.ac.in

Abstract. This paper addresses the issues of handling varying size images in convolutional neural networks (CNNs). When images of different size are given as input to a CNN then it results in varying size set of activation maps at its convolution layer. We propose to explore two approaches to address varying size set of activation maps for the classification task. In the first approach, we explore deep spatial pyramid match kernel (DSPMK) to compute a matching score between two varying size sets of activation maps. We also propose to explore different pooling and normalization techniques for computing DSPMK. In the second approach, we propose to use spatial pyramid pooling (SPP) layer in CNN architectures to remove fixed-length constraint and to allow the original varying size image as input to train and fine-tune the CNN for different datasets. Experimental results show that proposed DSPMK-based SVM and SPP-layer based CNN frameworks achieve state-of-the-art results for scene image classification and fine-grained bird species classification tasks.

Keywords: Convolutional neural network
Deep spatial pyramid match kernel · Image classification
Varying size set of activation map · Spatial pyramid pooling layer
Support vector machine

1 Introduction

Nowadays CNNs have been popular for their relevance to wide extent of applications, such as image segmentation [22], object classification [4,12,31], scene image classification [18,41], fine-grained classification [2,9,43] and so on. Fine-grained recognition has recently become popular [2,44], because it is applicable in a variety of challenging domain such as bird species recognition [35] or flower species recognition [27]. An important issue in fine-grained recognition is

M. De Marsico et al. (Eds.): ICPRAM 2018, LNCS 11351, pp. 44–64, 2019.
https://doi.org/10.1007/978-3-030-05499-1_3

inter-class similarity *i.e*, images of birds with different species can be ambiguous due to uncontrolled natural settings. On the other hand, generic scene image recognition is challenging task because scene images are composed with spatially correlated layout of different objects and concepts [19]. Successful recognition methods need to extract powerful visual representations to deal with high intra-class and low inter-class variability [38], complex semantic structure [29], varying size of same semantic concept across dataset, and so on. For addressing such issues many CNNs like, AlexNet [23], GoogLeNet [32] and VGGNet-16 [31] have already been trained on datasets like Places [45] and ImageNet [7] for image recognition tasks. These deep networks can be altered and prepare to train for other datasets and applications with less modifications. In all similar scenarios, features acquired from pre-trained, altered or fine-tuned CNNs are used to build standard classifier like fully connected neural network or support vector machine (SVM).

One major drawback of these frameworks is that the CNNs require the input images to be of fixed dimensions. For instance, GoogLeNet accepts images of resolution "224 × 224". Although the standard datasets like SUN397 [38] or MIT-67 [29] consist of variable resolution images which are much bigger than "224 × 224". Similarly, in case of CUB-200-2011 [35] bird dataset images are varying in size. Also, as we demonstrate, it is useful to consider bird region of interest (ROI) which focuses on the subject, and discards most of the background. In such cases, too the size of ROI can vary with the shape and size of the birds. The traditional methods to use these CNNs is to reshape the random-sized images to a same size. This leads to dissipation of information of the images before giving as input to the CNN for extracting the feature. The capability of classifier to give better results gets affected due to such usage, which is evident from the work published in [18]. To avoid any such prior information loss, different approaches are explored to feed varying resolution images as input to CNN. The works in [18], eliminates the necessity of fixed resolution image by including a spatial pyramid pooling (SPP) layer in CNN and titled the new architecture as SPP-Net. The works in [11], follows the similar technique by evaluating the feature maps of conv layer into a super vector using one of the encoding like Fisher vector (FV) [41] or vector of locally aggregated descriptor (VLAD) [20] by building the Gaussian mixture model (GMM).

As conv layers are the necessary part of convolutional neural network and responsible for producing discriminative activation maps. Generated activation maps are of varying resolution according to original image size and contain more spatial layout information compared to the activation of the fc layers, as fc layer integrates the spatial content present in the conv layer features. Inspired by the same fact, in our previously published work [16], we focused on passing the images in their actual size as input to the convolutional neural network and then acquire varying size sets of deep activation maps from the last conv layer as output.

In literature study, mainly two approaches are proposed to handle varying size pattern classification using support vector machines. In the first, a varying size set of activation maps is transformed into a fixed dimension pattern as in [11], and further a kernel function for fixed dimension pattern is used to build the support vector machine classifier. In the second, a suitable kernel function is directly designed for varying size set of activation maps. The kernel designed for varying size set of features is called dynamic kernels [8]. The dynamic kernels in [8,15,17,24] shows promising results for classification of varying resolution images and speech signals. We adopt the second approach and propose to design deep spatial pyramid match kernel (DSPMK) as dynamic kernel.

In this work, we extended the previous work of [16] and propose to explore different pooling and normalization techniques for computing DSPMK which is discussed in Sect. 3.2. Inspired from [18], we propose to consider the CNN architecture for fine-tuning by passing original images as input and added spatial pyramid pooling (SPP) layer to the network for handling the same. SPP-layer maps varying size activation maps to fixed size for passing to fully-connected layer. SPP-layer allows for end-to-end fine-tuning and training of the network with variable size images. This is discussed in Sect. 3.1. The key contribution of this work are:

- Deep spatial pyramid match kernel with different pooling and normalization technique to find the similarity score between a pair of varying size set of deep activation maps.
- Introducing SPP-layer [18] in between last convolutional layer and first fc-layer, so that varying size deep activation maps of images can be converted into fixed length representation.
- End-to-end fine-tuning of the network for different dataset with SPP-layer to handle the images in their original size.
- We demonstrate the effectiveness of our approach and its variants, with state-of-the-art results, over two different applications of scene image classification and fine-grained bird image classification.

The rest of the paper is structured as follows: A review of related approaches for image classification using CNN-based features is presented in Sect. 2. Section 3.1, gives the detail about CNN architecture with SPP-layer. In Sect. 3.2, we discuss the DSPMK for varying size set of deep activation maps with different pooling and normalization technique. The experimental studies using the proposed app-roach on scene image classification and fine-grained bird classification tasks is presented in Sect. 4. In Sect. 5 conclusion is presented.

2   Literature Review

In this section, we revisit the state-of-the-art techniques for fine-grained image classification and scene image classification tasks. Traditional method of image classification includes generating the local feature vector of images using local descriptors like, scale invariant feature transform (SIFT) [25] and histogram of oriented gradient (HOG) [6]. Further, GMM-based or SVM-based classifier can be built using the standard function such as Gaussian kernel, where the feature vectors are encoded into a fixed length representation. Generally bag of

Visual Words (BoVW) [5, 36, 37], sparse coding [40], and Fisher vector (FV) [26] encoding is used for fixed-dimensional representations of an image. These fixed length vector representations does not incorporate spatial information of concepts present in the image.

As an alternative, SVM-based classifiers can be learn with matching based dynamic kernels which are designed with consideration of spatial information. Spatial pyramid match kernel [24], class independent GMM-based intermediate matching kernel [8] and segment-level pyramid match kernel [15] are few of the matching based kernels for matching different size images and speech signals. With the development of deep CNNs, conventional features and related methods are being replaced by leaned features from datasets with linear kernel (LK) based SVM classifier.

The eye-popping performance of various deep CNN architectures on ImageNet large scale visual recognition challenge (ILSVRC) [23, 31, 32] has motivated the research community to adapt CNNs to other challenging domain and datasets like fine-grained classification. Initially, fc-layer features from convolutional neural network were directly in use to build SVM-based classifier using LK for any task of vision and perform batter than traditional methods [9, 46]. Few researchers also encoded learned features into a novel representation e.g, in [26] authors have transformed the features from fc layer to bag of semantics representation. This bag of semantic representation is then summaries in semantic Fisher vector representation. In case of fine-grained bird classification, state-of-the-art approaches are based on deep CNNs [2, 9, 39, 44]. These approach consider part based and bounding box annotation for generating the final representation. Moreover, all these approaches are based on giving fixed size input to the network because of rigid nature of fc-layer as it is based on fixed number of fully connected neurons and expects a fixed length representation of input, whereas the convolution process is not constrained with fixed length representation. So we can say, the necessity of fixed resolution image as input to convolutional networks is an mandatory demand of the fc-layer.

The impact of reshaping the images to a fixed size results in loss of information [18]. On the other side, convolutional layers of CNNs accept any arbitrary sized input image which results in random sized deep activation maps according to the input. Deep activations maps contain the strongest response of filters on the previous layer output and conserve the spatial information of the concepts present in input image. From the work in [11, 18], we can observe the similar idea. The approaches in these papers considered spatial pyramid based approach and scaled space of input images to incorporate the concept information of images into the activation maps at different scales. The work in [42], focuses on scale characteristics of images over feature activations. They consider images at different scales to input to the CNN and obtain seven layer pyramid of dense activation maps. Further they have used Fisher framework for encoding the activation maps to aggregate into a fixed length representation. The work proposed in [18], considers the SPP approach to expel the essentialness of same size image as input to convolutional networks. Here, the CNN is fed with images of original size. However, in [42] the CNN is fed with differently scaled images. The work in [11], also follow the similar way and fed the original sized image to convolutional network. However, the approach for converting into fixed size is different. Here, a GMM using fixed size vector representation obtained from spatial pyramid pooling, is built to generate the Fisher vectors [11]. Finally, all the Fishers vectors are concatenated to form a fixed dimensional representation.

In our work, we focuses on integrating the power convolutional-based varying size set of deep activation maps with dynamic kernel to obtain a matching value between a pair of images of different size. We used DSPMK as dynamic kernel rather of building GMM based dictionary on varying size conv features. In this way, our proposed approach is computationally less expensive. Further we modify the deep CNN architecture by adding the spatial pyramid pooling for end-to-end fine-tuning or training. In the next Section, we discuss CNN architecture with SPP-layer and the proposed DSPMK for the varying size set of deep activation maps.

3   Approaches for Handling Variable Size Images for Classification

In this section, we discuss the approaches to handle the variable size images in CNN for classification on different domain datasets like scene image classification dataset and fine-grained bird classification dataset.

- In Sect. 3.1, we introduce SPP-layer inspired by [18] in between last convolutional layer and first dense layer so that variable size set of convolutional activation maps of images can be converted into fixed length representation for end-to-end training of the network.
- In Sect. 3.2, we present DSPMK proposed in [16] to compute the similarity score between two sample images represented as varying size set of activation maps.

3.1   CNN Architecture with Spatial Pyramid Pooling (SPP) Layer

As mentioned earlier, traditional CNN architecture like AlexNet [23], GoogLeNet [32] or VGGNet-16 [31] are pre-trained on the dataset with images of fixed resolution (e.g 227 × 227). Conversion of image from original size to fixed size results in loss of information in the beginning of network. CNN architecture is mainly the combination of convolutional (conv) layer and fully-connected (fc) layer. The fc-layer demands fixed size input and conv-layers are free from such restrictions. In this work, we modify the CNN architecture such that images are allowed in its original size for input and gives varying size set of activation maps as output from last convolutional layer.

To handle this further, we propose to use spatial pyramid pooling (SPP) layer inspired by [18] which map varying size set of deep activation maps onto a fixed length representation for end-to-end training. Using SPP-layer, information aggregation happens at later stage in the network which improve the training process. We have considered VGG-19 architecture [30,31] and added SPP-layer between last convolutional layer and first fc-layer. The SPP-layer sum-pools the varying length convolutional layer activation maps at three different levels to convert them into fixed length vector. In the first level complete convolutional activation maps are considered and sum or max-pool is applied to obtain a fixed length vector. In the second level, activation maps are spatially divided into 4 blocks and sum or max-pooling is applied in respective block for converting the variable size deep activation maps to fixed length vector. In the third level, activation maps are divided into 16 blocks and so on. In this scenario we are fixing the number of spatially divided blocks instead of block size. The fixed length vectors obtained in each level are concatenated to form a fixed length supervector. This fixed length supervector is further passed onto fully-connected layer for end-to-end training using back propagation. The block diagram of proposed CNN architecture with SPP-layer is shown in Fig. 1.

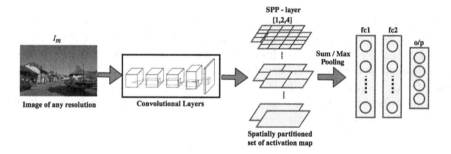

Fig. 1. Block diagram of CNN architecture with SPP-layer.

3.2   Deep Spatial Pyramid Match Kernel

In this section, we present deep spatial pyramid match kernel (DSPMK) proposed in [16] for matching varying size set of deep activation maps obtained from convolutional neural network. The entire process of classification using DSPMK-based SVM is demonstrated in block diagram of Fig. 2. As shown in diagram, I_m and I_n are two images given to the convolutional layer of network as input such that we get set of deep activation maps. Different image give variant size activation maps as output i.e, activation maps in set corresponding to image I_m is different from image I_n. From these different size activation maps, we propose to compute similarity sore using DSPMK. DSPMK-based SVM classifier is learn by association of feature maps of training images with the class label. This is in contrast to [11], In [11] varying size activation maps are transformed into fixed size using Fisher framework and are encoded to fixed length super vector like Fisher vector and then LK-based SVM is used for building the classifier. Main features of the proposed approach is that, DSPMK computes the similarity score on different size actual images at different spatial levels ranging from 0, 1 to L using varying size set of deep activation maps.

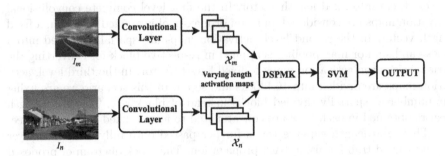

Fig. 2. Block diagram of DSPMK as proposed in [16].

Consider dataset of images as $\mathcal{D} = \{I_1, I_2, \ldots, I_m, \ldots, I_N\}$ and 'f' be the number of kernels or filters in last conv layer of pre-defined deep CNN architecture. Let the mapping $\mathcal{F}$, takes input, actual image and project it to set of deep activation maps using conv layers of CNN. Mapping $\mathcal{F}$ is given as, $\mathcal{X}_m = \mathcal{F}(I_m)$. Size of activation maps obtain from last conv layer in a set corresponding to a image is same but vary from image to image as images are fed in its original resolution to the CNN architecture.

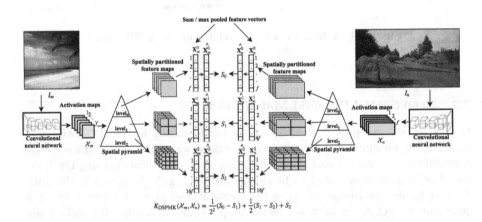

$$K_{\text{DSPMK}}(\mathcal{X}_m, \mathcal{X}_n) = \frac{1}{2^2}(S_0 - S_1) + \frac{1}{2}(S_1 - S_2) + S_2$$

Fig. 3. Illustration of computing similarity score using DSPMK between two different resolution images I_m and I_n, similar to the Fig. 2 of paper [16]. Here, $\mathcal{X}_m$ and $\mathcal{X}_n$ are set of deep activation map computed using conv layer of pre-trained CNN, size of $\mathcal{X}_m$ depends on size of I_m, similarly size of $\mathcal{X}_n$ depends on size of I_n. The matching score at each level l (*i.e*, S_0, S_1 and S_2) is computed using Eq. (3).

Firstly, images to pre-trained CNN is fed in its actual size. For image I_n, we have a set $\mathcal{X}_n = \{\mathbf{x}_{n1}, \mathbf{x}_{n2}, \mathbf{x}_{n3}, \dots, \mathbf{x}_{nf}\}$ consisting of 'f' feature maps from mapping $\mathcal{F}$, where $\mathbf{x}_{ni} \in \mathbb{R}^{p_n \times q_n}$ and $p_n \times q_n$ is the size of each feature map obtained from last conv layer which varies accordant to the input image resolution. This conclude to varying size deep activation map as shown in Fig. 3 for images I_m and I_n.

Secondly, deep activation maps are spatially divided into sub-blocks to form spatial pyramid. At level-0, activation maps are considered as it is without spatial division. At level-1, every deep activation map is divided into 4 spatial divisions related to 4 quadrant, as depicted in Fig. 3. Consider $L + 1$ the total number of levels in the pyramid start from 0, 1 till L. In At any level-l, a deep activation map $\mathbf{x}_{ni}$ is spatially split into 2^{2l} blocks. At any level-l, activation values of cells in every spatial block of all the f deep activation maps are sum or max-pooled and concatenated to form a vector $\mathbf{X}_n^l$ of size $f2^{2l} \times 1$. This scenario is expatiated in Fig. 3 by considering three different levels, $l = 0$, 1 and 2 and same is also described in Algorithm 1 for $L + 1$ pyramid levels.

In our proposed framework, we considered three spatial pyramid levels. At level-0, ($i.e$, $l = 0$) the complete activation maps corresponding to input image is sum or max-pooled, in total there are f activation maps in output of conv layer which results is $f \times 1$ size vector representation. At level-1 ($i.e$, $l = 1$), the same activation maps are considered again and divided into four equal spatial blocks. Each block correspond to single activation maps are again sum or max pooled, which results in 1×4 size vector. Same procedure is repeated for f activation maps which results into a vector of $4f \times 1$ size. Similarly, at level-2 ($i.e.$ $l = 2$), again the same activation maps are divided into sixteen equal spatial regions resulting into a vector of $16f \times 1$ dimensional vector. Corresponding to image I_n, after concatenating all the sum or max pooled activation values are results in a single vector called $\mathbf{X}_n^l$.

The $\mathbf{X}_m^l$ can now be seen as representation of image I_m at level-l of pyramid. At this stage, we propose to compute deep spatial pyramid match kernel (DSPMK) to match two images rather than deriving Fisher vector (FV) representation as in [11]. Our proposed approach avoids building GMM to obtain FV and hence reduces the computation complexity as compared to [11]. The process of computing DSPMK is motivated from spatial pyramid match kernel (SPMK) [24]. SPMK involves the histogram intersection function that match the frequency based image representation or normalized vector representation of two images at every levels of pyramid [24]. However, $\mathbf{X}_m^l$ is not in the normalized vector representation of image I_m. We propose to normalized $\mathbf{X}_m^l$ using ℓ_1 and ℓ_2 to obtain normalized vector representation.

Let $\mathbf{X}_m^l$ and $\mathbf{X}_n^l$ be the representation at level-l corresponding to two images I_m and I_n respectively. The normalized vector representation of $\mathbf{X}_m^l$ and $\mathbf{X}_n^l$ is obtained using ℓ_1 or ℓ_2 normalization as given in Eqs. (1) and (2)

$$\widehat{\mathbf{X}}_m^l = \frac{\mathbf{X}_m^l}{||\mathbf{X}_m^l||_1}, \widehat{\mathbf{X}}_n^l = \frac{\mathbf{X}_n^l}{||\mathbf{X}_n^l||_1} \tag{1}$$

Algorithm 1. Deep spatial pyramid matching kernel $K_{\text{DSPMK}}(\mathcal{X}_m, \mathcal{X}_n)$.

Require:

(i) Activation maps set $\mathcal{X}_m$ and $\mathcal{X}_n$

    $\mathcal{X}_m = \{\mathbf{x}_{m1}, ..., \mathbf{x}_{mi}, ..., \mathbf{x}_{mf}\}$; where $\mathbf{x}_{mi} \in \mathbb{R}^{p_m \times q_m}$

    $\mathcal{X}_n = \{\mathbf{x}_{n1}, ..., \mathbf{x}_{ni}, ..., \mathbf{x}_{nf}\}$; where $\mathbf{x}_{mi} \in \mathbb{R}^{p_n \times q_n}$

(ii) $L + 1$: number of pyramid levels.

1: **Procedure:**

2: **for** $l = 0$ **to** L **do**

3:    Divide each activation map of $\mathcal{X}_m$ into 2^{2l} equal spatial blocks.

    $\mathcal{X}_m^l =$

    $\{\mathbf{x}_{m1(1)}^l ... \mathbf{x}_{m1(2^{2l})}^l, ..., \mathbf{x}_{mi(1)}^l ... \mathbf{x}_{mi(2^{2l})}^l, ..., \mathbf{x}_{mf(1)}^l ... \mathbf{x}_{mf(2^{2l})}^l\}$

4:    Apply sum or max-pooling over each block such that

    $x_{mi(j)}^l = \sum_u \sum_v \mathbf{x}_{mi(j)}^l(u, v)$

    $\mathbf{X}_m^l =$

    $\{x_{m1(1)}^l ... x_{m1(2^{2l})}^l, ..., x_{mi(1)}^l ... x_{mi(2^{2l})}^l, ..., x_{mf(1)}^l ... x_{mf(2^{2l})}^l\}$

    $\in \mathbb{R}^{f2^{2l} \times 1}$

5:    Normalize the generated feature vector $\mathbf{X}_m^l$ using ℓ_1 or ℓ_2 norm

    $\widehat{\mathbf{X}}_m^l =$

    $\{\hat{x}_{m1(1)}^l ... \hat{x}_{m1(2^{2l})}^l, ..., \hat{x}_{mi(1)}^l ... \hat{x}_{mi(2^{2l})}^l, ..., \hat{x}_{mf(1)}^l ... \hat{x}_{mf(2^{2l})}^l\}$

    $\in \mathbb{R}^{2^{2l} \times f \times 1}$

6:    Repeat step 3 to 5 for computing $\widehat{\mathbf{X}}_n^l$ for image I_n

7:    Compute intermediate similarity score S_l between $\widehat{\mathbf{X}}_m^l$ and $\widehat{\mathbf{X}}_n^l$ using Equation (3).

8: **end for**

9: Compute final similarity score between $\mathcal{X}_m$ and $\mathcal{X}_n$ using Equation (4).

Ensure:

9: $K_{\text{DSPMK}}(\mathcal{X}_m, \mathcal{X}_n)$

$$\widehat{\mathbf{X}}_m^l = \frac{\mathbf{X}_m^l}{||\mathbf{X}_m^l||_2}, \widehat{\mathbf{X}}_n^l = \frac{\mathbf{X}_n^l}{||\mathbf{X}_n^l||_2} \tag{2}$$

The Histogram intersection (HI) function is used to compute intermediate matching score S_l between $\widehat{\mathbf{X}}_m^l$ and $\widehat{\mathbf{X}}_n^l$ at each level l as,

$$S_l = \sum_{j=1}^{f} \sum_{k=1}^{2^{2l}} min(\hat{x}_{mj(k)}^l, \hat{x}_{nj(k)}^l) \tag{3}$$

Here, the intermediate similarity score S_l found at level-l also includes all the matches found at the finer level $l + 1$. As a result, the number of new matches found at level l is given by $S_l - S_{l+1}$ for $l = 0, ..., L - 1$. The DSPMK is computed as a weighted sum of the number of new matches at different levels of the spatial pyramid. The weight associated with level l is set to $\frac{1}{2^{(L-l)}}$, which is inversely proportional to width of spatial regions at that level.

The DSPMK kernel is computed as,

$$K_{\text{DSPMK}}(\mathcal{X}_m, \mathcal{X}_n) = \sum_{l=0}^{L-1} \frac{1}{2^{L-l}}(S_l - S_{l+1}) + S_L \qquad (4)$$

The main advantages of proposed approach is that it incorporate any size image without any resizing loss and it combines the convolutional varying size deep activation maps with dynamic kernel named DSPMK based SVM.

4 Experimental Studies

In this section, the efficacy of the proposed framework is studied on scene image and bird species classification task using SVM-based classifier. In experiments, we cover mainly two aspects for handling varying nature of image; one by computing the varying size deep activation maps from last convolutional layer and compute the classification score using DSPMK, and the other by adding the spatial pyramid pooling layer to the network for handling varying nature of image and fine-tuned it with respective dataset for computing the fully trained features.

4.1 Datasets

We tested our proposed approach on two different kinds of datasets one for scene classification which includes datasets such as MIT-8 Scene [28], Vogel-Schiele (VS) [34], MIT-67 [29] and SUN-397 [38], and the other for fine-grained bird species classification with the CUB-200-2011 [35] dataset.

MIT-8-scene: This dataset contain total of 2688 scene images belonging to 8 different semantic classes, like, 'coast', 'mountain', 'forest', 'open-country', 'inside-city', 'highway', 'tall building' and 'street'. We randomly select 100 scene images from each class for training the model and keep remaining images for testing. We consider 5 such sets. The final classification scores computed in this paper correspond to the average classification accuracy for 5 trials.

Vogel-Schiele: This dataset contain total of 700 scene images belonging to 6 different semantic classes, viz., 'forests', 'mountains', 'coasts', 'river', 'open-country', and 'sky-clouds'. We consider 5-fold stratified cross validation and present the result as average classification score of 5-fold.

MIT-67: This is indoor scene dataset. Most of the scene recognition models work well for outdoor scenes but perform poorly in the indoor domain. This dataset contain 15,620 images with 67 scene categories. All images have a minimum resolution of 200 pixels in the smallest axis. It is a challenging dataset, due to the less in-class variability. The standard division [29] for this dataset consist of approximately 80 images of each class for training and 20 images for testing.

SUN397: This database contains 397 categories used in the benchmark of several paper. The number of images varies across categories like indoor, urban and nature but there are at least 100 images per category, and 108,754 images in total. We consider publicly available fixed train and test splits from [38], where each split has 50 training and 50 testing images per category. We consider the first five split set and the result computed is the average classification accuracy for 5 splits.

Caltech-UCSD Birds CUB-200-2011 dataset consists of 11,788 images of birds belonging to 200 different species. The standard division for this dataset consist of 5994 images for training and 5794 for testing [35] with approximately 30 images per species in the training class, and the rest in test class. Bird data suffers from high intra-class and low inter-class variance. Dataset is available with bird bounding boxes and other annotations. In this work, we evaluate our methods in two scenarios one with the bounding-box which enables one to focus on the bird region rather then background and other without bounding-box information considered at training and test time.

4.2   Experiment Studies for Scene Image Classification Task

In our studies of scene image classification, we have consider different CNN architectures for extracting the features like, AlexNet [23], GoogLeNet [32] and VGGNet-16 [31] which are pre-trained on three different datasets, i.e, ImageNet [7], Places205 and Places365 [45] datasets. Reason behind using the different pre-trained networks (on different datasets) is that all used datasets consist of variety of images. In this context, ImageNet dataset contains mainly object centric images and it shows activations for object-like structures, whereas Places dataset comprise of largely indoor/outdoor scene images. We believe that CNNs trained on Places dataset activate for landscapes, natural structure of scenes with more spatial features, and indoor scene patterns.

In all the convolutional networks, pre-trained weights are kept consistent without fine-tuning. These networks are in use without its fc-layers in our experimental studies so that input images of arbitrary size can be accepted. As discussed in Sect. 3.2, we have passed the original image of arbitrary size as input to deep CNNs and extracted varying size set of deep activation maps from last convolutional layer. The size of set of activation map corresponding to an image depends on the filter size, number of filters f in last convolutional layer and input image size. The number of filters f, in last convolution layer of AlexNet, GoogLeNet and VGGNet-16 are 256, 1024 and 512 respectively. The architecture of these CNNs also differs from each other. So, activation map size will vary from image to image and architecture to architecture.

DSPMK between varying size deep activation map for pair of images is computed as in Fig. 3 using Eq. (1) to (4). We consider $L + 1 = 3$ as the number of levels in spatial pyramid. In computation of DSPMK, we have performed the experiments with both sum and max-pooling techniques. Reason behind using different pooling technique is, max-pooling extracts the most activated feature like edges, corner and texture, whereas, sum-pooling smoothen out the activation map and measures the sum value of existence of a pattern in a given region. Although results with both the pooling technique are comparable as shown in Tables 1 and 2, we observed that max-pooling works a bit better then sumpooling in our case. It is seen that performance of SVM-based classifier with

Table 1. Comparison of classification accuracy (CA) (in %) with 95% confidence interval for the **SVM-based classifier** using DSPMK computed using sum-pooling on different datasets, similar to study shown in Table 1 of paper [16]. Base features for the proposed approach are extracted from different CNN architecture like, AlexNet, GoogLeNet and VGGNet which are pre-trained deep network on ImageNet, Places-365 and Places-205 dataset respectively.

Different pre-trained deep CNN architectures used to build DSPMK with sum-pooling	MIT-8 scene	Vogel-Schiele	MIT-67	SUN-397
ImageNet-AlexNet [23]	93.52±0.13	79.46±0.23	62.46	45.46±0.12
Places205-AlexNet [46]	93.56±0.12	82.21±0.25	62.24	53.21±0.23
Places365-AlexNet [45]	94.15±0.11	82.90±0.31	66.67	55.43±0.24
ImageNet-GoogLeNet [32]	92.02±0.06	82.30±0.25	71.78	50.32±0.31
Places205-GoogLeNet [46]	92.15±0.18	85.84±0.36	75.97	57.43±0.26
Places365-GoogLeNet [45]	93.70±0.16	85.54±0.21	75.60	59.89±0.21
ImageNet-VGG [31]	93.90±0.07	84.62±0.31	75.78	53.67±0.25
Places205-VGG [46]	94.54±0.03	**86.92±0.26**	**81.87**	61.86±0.24
Places365-VGG [45]	**95.09±0.14**	84.68±0.28	77.76	**62.31±0.25**

DSPMK obtained using deep features from VGGNet-16 is significantly better than that of SVM with DSPMK obtained using deep features from GoogLeNet and AlexNet. Reason being VGGNet-16 has very deep network compare to other architectures and it learns the hierarchical representation of visual data more efficiently. We consider LIBSVM [3] tool to build the DSPMK-based SVM classifier. Specifically, we uses one-against-the-rest approach for multi-class scene image classification. In SVM for building the classifier, we use default value of trade-off parameter $C = 1$. In our further study, we fine-tuned the VGG-16 architecture for respective datasets by adding the spatial pyramid pooling (SPP) layer to the network as shown in Fig. 2. We computed the spatial pyramid pooling features and train the neural network based classifier. We consider the neural network with two hidden layer and one soft-max layer. Dropout is chosen as 0.5 learning rate as 0.01 and 2048 neurons in the hidden layers. We observe that results are comparable with DSPMK-based SVM approach.

Table 3 presents the comparison of scene image classification accuracy of proposed DSPMK-based SVM classifier and the SPP-based neural network classifier with that of state-of-the-art approaches. From Table 3, it is seen that both of our proposed approaches are giving better performance in comparison with

Table 2. Comparison of classification accuracy (CA) (in %) with 95% confidence interval for the **SVM-based classifier** using DSPMK computed using max-pooling on different datasets, similar to study shown in Table 1 of paper [16]. Base features for the proposed approach are extracted from different CNN architecture like, AlexNet, GoogLeNet and VGGNet which are pre-trained deep network on ImageNet, Places-365 and Places-205 dataset respectively.

Different pre-trained deep CNN architectures used to build DSPMK with max-pooling	MIT-8 scene	Vogel-Schiele	MIT-67	SUN-397
ImageNet-AlexNet [23]	94.12±0.11	80.17±0.16	63.67	46.12±0.13
Places205-AlexNet [46]	94.11±0.13	83.18±0.21	63.56	54.01±0.21
Places365-AlexNet [45]	94.65±0.08	83.11±0.20	68.21	56.12±0.23
ImageNet-GoogLeNet [32]	91.12±0.09	83.21±0.27	72.99	52.12±0.28
Places205-GoogLeNet [46]	93.14±0.12	86.91±0.31	76.82	56.12±0.24
Places365-GoogLeNet [45]	92.89±0.13	86.67±0.24	77.22	60.13±0.23
ImageNet-VGG [31]	93.86±0.11	85.21±0.33	75.99	54.91±0.22
Places205-VGG [46]	95.56±0.06	**87.81±0.21**	**82.83**	62.76±0.22
Places365-VGG [45]	**96.21±0.09**	85.66±0.30	78.16	**63.12±0.21**

traditional feature based approaches in [21,25] and also with CNN-based approaches in [11,13,26,42,46].

The works in [25], uses scale invariant feature transform (SIFT) descriptors to represent images as set of local feature vectors, which are further converted into bag-of-visual word (BOVW) representation for classification using linear kernel based SVM classifier. The works in [21] uses the learned bag-of-part (BoP) representation and combine with improved Fisher vector for building linear kernel based SVM classifier. The works in [13] extracted CNN-based features from multiple scale of image at different levels and performs orderless vectors of locally aggregated descriptors (VLAD) pooling [20] at every scale separately. The representations from different level are then concatenated to form a new representation known as multi-scale orderless pooling (MOP) which is used for training linear kernel based SVM classifier. The works in [46] uses more direct approach, where a large scale image dataset (Places dataset) is used for training the AlexNet architecture and extracted fully-connected (fc7) layer feature from the trained network. The basic architecture of their Places-CNN is same as that of the AlexNet [23] trained on ImageNet. The works in [46] trained a Hybrid-CNN, by combining the training data of Places dataset with ImageNet dataset. Here, features from fully-connected (fc7) layer are then used for training linear kernel

Table 3. Comparison of classification accuracy (CA) (in %) with 95% confidence interval of proposed approach with state-of-the-art approaches on MIT-8 scene, Vogel-Schiele, MIT-67 Indoor and SUN-397 dataset, similar to study shown in Sect. 4, Table 2 of paper [16]. (SIFT: Scale invariant feature transform, IFK: Improved Fisher kernel, BoP: Bag of part, MOP: Multi-scale orderless pooling, FV: Fisher vector, DSP: Deep spatial pyramid, MPP: Multi-scale pyramid pooling, DSFL: Discriminative and shareable feature learning and NN: Neural network).

Method	MIT-8-Scene	Vogel Schiele	MIT-67	SUN-397
SIFT+BOVW [25]	79.13±0.13	67.49±0.21	45.86	24.82±0.34
IFK+BoP [21]	85.76±0.12	73.23±0.23	63.18	-
MOP-CNN [13]	89.45±0.11	76.81±0.27	68.88	51.98±0.24
Places-CNN-fc7 [46]	88.30±0.09	76.02±0.31	68.24	54.32±0.14
Hybrid-CNN-fc7 [46]	91.23±0.04	78.56 ±0.21	70.80	53.86±0.21
fc8-FV [26]	88.43±0.08	79.56±0.23	72.86	54.40±0.30
VGGNet-16 + DSP [11]	92.34±0.12	81.34±0.27	76.34	57.27±0.34
MPP(Alex-fc7)+DSFL [42]	93.21±0.14	82.12±0.25	80.78	-
VGG16 + SPP-feature + NN based classifier (Ours)	94.01±0.11	85.89±0.23	80.94	-
VGG16 + DSPMK-based SVM with max-pooling (Ours)	**96.21±0.09**	**87.81±0.21**	**82.83**	**63.12±0.21**

based SVM classifier. The works in [26] obtained the semantic Fisher vector (FV) using standard Gaussian mixture encoding for CNN-based feature. Further linear kernel based SVM classifier is build using semantic FV for classification of scene images. The works in [11] uses the generative model based approach to build a dictionary on top of CNN activation maps. A FV representation for different spatial region of activation map is then obtained from the dictionary. A power and l_2 normalization is applied on the combined FV from different spatial region. A linear kernel based SVM classifier is then used for scene classification. The works in [42] combine the features from fc7 layer of AlexNet (Alex-fc7) and their complementary features named discriminative and shareable feature learning (DSFL). DSFL learns discriminative and shareable filters with a target dataset. The final image representation is used with the linear kernel based SVM classifier for the scene classification task.

In contrast to all the above briefly explained approaches, our proposed approaches use the image of arbitrary size and gives the deep activation map of varying size without any loss of information. The deep spatial pyramid match kernel can handle the varying size set of deep activation maps and incorporates the local spatial information at the time of computing level wise matching score. Specifically, our proposed approach is very simple and discriminative in nature which outperforms the other CNN-based approaches without combining any complementary features as in [42]. Our first proposed approach, based on SPP-feature with neural network (NN) also shows good quality results (second to only our proposed DSPMK method), as this approach consider original size images for fine-tuning the network. Our second proposed framework, bring out that for scene recognition, good performance is accomplishable by using last conv layer features with DSPMK-based-SVM. Proposed framework is free of fully connected layer, believe on the actual size image, memory efficient, simple and take very less computing time in compare to state-of-the-art techniques.

4.3 Experiment Studies for Fine-Grained Bird Species Classification

The experiments for fine-grained bird species classification cover three main aspects of our approach. First, we compute varying size deep activation map by passing images in its original size without any prior loss of information. Second, we use DSPMK to compute matching score between them. Third, we fine-tune the VGG-19 architecture [30] by adding SPP-layer to it. We fine-tune the network for CUB-200-2011 dataset [35] and compute variable size deep activation map features and SPP-features for further experiments. We show our proposed approach is generic and along with scene image classification it works well for fine-grained bird species classification.

Table 4, shows the results of fine-grained bird species classification with different methods. We have shown the results for testing with bounding box (Bbox) and without bounding box. The bounding box annotation essentially helps us to crop only the prominent bird region of interest (RoI) while discarding the background. Such regions may also be obtained by detection algorithm. The case without Bbox corresponds to complete actual image. Firstly, we passes the image in fixed size i.e, "224 × 224" for both the cases to the CNN architecture and computed fixed length fc7 and pool5 features. We use linear kernel based SVM to compute the classification score. Secondly, we pass the image in its original size without resizing it to "224 × 224" and computed varying size deep activation maps. In this context, we perform experiments using DSPMK-based SVM with different pooling technique for computing the classification score. Next, we fine-tune the VGG-19 architecture by adding SPP-layer between last convolutional layer and first fully-connected layer. We consider the fine-tuned network for further experiments in two ways. In the first approach, we compute the varying size set of deep activation map and use DSPMK-based SVM for computation of classification score. In the second approach, we compute SPP-features from fine-tuned network and train neural network based classifier. In this context, we uses two hidden layer with 2096 neurons in each. We empirically chosen learning rate as 0.001 and dropout as 0.5.

We observe in Table 4 that, if the images are not resized and no Bbox RoI detection is available, original images can be used instead with proposed

Table 4. Comparison of classification accuracy (CA) (in %) for the SVM-based classifier using linear kernel and DSPMK, fine-tuned VGG19 with SPP-layer based neural network on CUB-200-2011 dataset. Proposed approach uses base features extracted from VGG19 [30]. Here NN indicate neural network.

Method	Testing with Bbox	Testing without Bbox
VGG19-fc7+ linear kernel based SVM	79.02	72.94
VGG19-Pool5+ linear kernel based SVM	78.30	69.84
SVM using DSPMK with sum-pooling	82.07	74.36
SVM using DSPMK with max-pooling	**82.12**	80.81
VGG19 + fine-tuning with SPP + fc7	78.41	76.63
VGG19 + fine-tuning with SPP + SVM using DSPMK with sum-pooling	79.11	78.16
VGG19 + fine-tuning with SPP + SVM using DSPMK with max-pooling	80.01	78.89
VGG19 + SPP-feature + NN based classifier	81.24	**81.03**

DSPMK-based SVM approach. In this context, one can notice that classification accuracy will be marginally affected. This is natural as the case with Bbox focuses only on the bird RoI. However, this difference is relatively small for most of variants of the proposed methods using DSPMK and SPP. This indicates that for bird images of the size and scale as in the CUB dataset, the proposed methods are largely invariant to ROI selection, and thus can obviate an ROI detection step. When images are used without bounding box annotation, we observe that there is huge i.e, (approx 10%) improvement in performance from linear kernel based SVM with VGG-19 pool5 features to DSPMK-based SVM with varying size activation maps features from last conv layer. We believe that, our proposed approach compute the matching score between two images more efficiently with consideration of spatial information.

In Table 5, we compare the classification results of proposed approaches with state-of-the-art results. The deformable part descriptor (DPD) in [44], is based on the supervised version of deformable part models (DPD) [10] for training, which then allows for pose normalization by comparing corresponding parts. The work in [1], learns a linear classifier for each pair of parts and classes.

Table 5. Comparison of classification accuracy (CA) (in %) on CUB-200-2011 dataset between different state-of-the-art method with that of the proposed approaches. Some of the state-of-the-art approaches uses part annotations during training and testing. The proposed approaches do not use any part information. (DPD: Deformable part descriptors; POOFs: Part-based One-vs-One Features; NN: Neural Network)

Method	Accuracy	Remark
DPD [44]	50.98	Uses parts info
POOFs [1]	56.78	Uses parts and Bbox info
Part transfer [14]	57.84	Uses parts and Bbox info
DeCAF$_6$ [9]	58.75	Uses Bbox info
DPD + DeCAF$_6$ [9]	64.96	Uses parts and Bbox info
Pose Normalized CNN [2]	75.70	Uses parts info
Parts-RCNN-FT [43]	76.37	Uses parts info
VGG19 + fine-tuning with SPP + fc7 (Ours)	78.41	Uses Bbox info
VGG19 + fine-tuning with SPP + DSPMK (Ours)	79.11	Uses Bbox info
VGG19 + SPP-feature + NN based classifier (Ours)	81.24	Uses Bbox info
DSPMK-based SVM with max-pooling (Ours)	**82.12**	Uses Bbox info

The decision values from many of such classifiers are used as feature representation. This approach also require ground-truth part annotations at training and also at test time. The work in, [14], is based on nonparametric part detection. Here, the basic idea is to use nearest neighbor matching to obtain similar training example from human-annotated part positions. The work in [9] is based on feature extraction from part regions detected using a DPM, which have sufficient depictive power and generalization ability to perform desired task. The work in [2] uses deep CNNs for extracting the features from image patches that are located and normalized by the pose. The work in [43], generate object proposals using Selective Search [33] and uses the part locations to calculate localized features from R-CNNs.

From Table 5, we also infer that our approaches for bird species classification does not require part annotation, and yet improves over very complex state-of-the-art approaches that use part based annotation at the time of training and testing. In contrast, our approaches are generic and easy to adapt to other datasets as we only require a pre-trained CNN architecture. For fine-tuning the CNN architecture with SPP-layer, we perform experiments without bounding box as well as with bounding box. It is observed that proposed framework perform much improved without any extra annotations.

5   Conclusion

In this work, we propose deep spatial pyramid match kernel (DSPMK) for improving the base features from last conv CNN's layer. DSPMK-based SVM can classify different size images which are represented as the varying size set of deep activation maps. Further, we propose to add spatial pyramid pooling layer in CNN architecture so that, we can fine-tune the pre-trained CNNs for other datasets containing varying size images. Our model has a dynamic kernel which calculates the layer-wise intermediate matching score and strengthens the matching procedure of conv layer features. The training of DSPMK-based SVM classifier take very less time in compare to training of GMM in [11]. In our research, we have considered the last convolutional layer features rather than fc layer features as fc layer limits these features to the fixed size and requires much larger computation time as it contains approximately 90% of the aggregate parameters of CNN. Thus, conv layer features are effectively considered in handling large varying size images in scene image classification datasets like, SUN-397 and MIT-67, as well as for size variations in the fine-grained classification with the CUB dataset. Almost all approaches in fine-grained classification are specialized, but we show that our approach is generic and works well for both the diverse datasets. In terms of performance, our proposed approach accomplishes state-of-the-art results for standard scene classification and bird species classification dataset. In future, for capturing differences of the activations caused by the varying size of concepts in an image, multi-scale deep spatial pyramid match kernel can be investigated.

References

1. Berg, T., Belhumeur, P.N.: Poof: part-based one-vs.-one features for fine-grained categorization, face verification, and attribute estimation. In: 2013 IEEE Conference on Computer Vision and Pattern Recognition (CVPR), pp. 955–962. IEEE (2013)
2. Branson, S., Van Horn, G., Belongie, S., Perona, P.: Bird species categorization using pose normalized deep convolutional nets. arXiv preprint arXiv:1406.2952 (2014)
3. Chang, C.C., Lin, C.J.: LIBSVM: A library for support vector machines. ACM Trans. Intell. Syst. Technol. 2, 27:1–27:27 (2011)
4. Chatfield, K., Simonyan, K., Vedaldi, A., Zisserman, A.: Return of the devil in the details: delving deep into convolutional nets. arXiv preprint arXiv:1405.3531 (2014)

5. Csurka, G., Dance, C., Fan, L., Willamowski, J., Bray, C.: Visual categorization with bags of keypoints. In: Workshop on Statistical Learning in Computer Vision, ECCV, vol. 1, pp. 1–2. Prague (2004)
6. Dalal, N., Triggs, B.: Histograms of oriented gradients for human detection. In: 2005 IEEE Conference on Computer Vision and Pattern Recognition (CVPR), vol. 1, pp. 886–893 (2005)
7. Deng, J., Dong, W., Socher, R., Li, L.J., Li, K., Fei-Fei, L.: Imagenet: a large-scale hierarchical image database. In: IEEE Conference on Computer Vision and Pattern Recognition, CVPR 2009, pp. 248–255. IEEE (2009)
8. Dileep, A.D., Chandra Sekhar, C.: GMM-based intermediate matching kernel for classification of varying length patterns of long duration speech using support vector machines. IEEE Trans. Neural Netw. Learn. Syst. **25**(8), 1421–1432 (2014)
9. Donahue, J., et al.: Decaf: a deep convolutional activation feature for generic visual recognition. In: International Conference on Machine Learning, pp. 647–655 (2014)
10. Felzenszwalb, P.F., Girshick, R.B., McAllester, D., Ramanan, D.: Object detection with discriminatively trained part-based models. IEEE Trans. Pattern Anal. Mach. Intell. **32**(9), 1627–1645 (2010)
11. Gao, B.B., Wei, X.S., Wu, J., Lin, W.: Deep spatial pyramid: the devil is once again in the details. CoRR abs/1504.05277 (2015)
12. Girshick, R., Donahue, J., Darrell, T., Malik, J.: Rich feature hierarchies for accurate object detection and semantic segmentation. In: Proceedings of the IEEE Conference on Computer Vision and Pattern Recognition, pp. 580–587 (2014)
13. Gong, Y., Wang, L., Guo, R., Lazebnik, S.: Multi-scale orderless pooling of deep convolutional activation features. In: Fleet, D., Pajdla, T., Schiele, B., Tuytelaars, T. (eds.) ECCV 2014. LNCS, vol. 8695, pp. 392–407. Springer, Cham (2014). https://doi.org/10.1007/978-3-319-10584-0_26
14. Göring, C., Rodner, E., Freytag, A., Denzler, J.: Nonparametric part transfer for fine-grained recognition. In: CVPR, vol. 1, p. 7 (2014)
15. Gupta, S., Dileep, A.D., Thenkanidiyoor, V.: Segment-level pyramid match kernels for the classification of varying length patterns of speech using svms. In: 2016 24th European Signal Processing Conference (EUSIPCO), pp. 2030–2034. IEEE (2016)
16. Gupta, S., Pradhan, D., Dileep, A.D., Thenkanidiyoor, V.: Deep spatial pyramid match kernel for scene classification. In: ICPRAM, pp. 141–148 (2018)
17. Gupta, S., Thenkanidiyoor, V., Aroor Dinesh, D.: Segment-level probabilistic sequence kernel based support vector machines for classification of varying length patterns of speech. In: Hirose, A., Ozawa, S., Doya, K., Ikeda, K., Lee, M., Liu, D. (eds.) ICONIP 2016. LNCS, vol. 9950, pp. 321–328. Springer, Cham (2016). https://doi.org/10.1007/978-3-319-46681-1_39
18. He, K., Zhang, X., Ren, S., Sun, J.: Spatial pyramid pooling in deep convolutional networks for visual recognition. IEEE Trans. Pattern Anal. Mach. Intell. **37**(9), 1904–1916 (2015)
19. Henderson, J.: Introduction to real-world scene perception. Vis. Cogn. **12**(6), 849–851 (2005)
20. Jégou, H., Douze, M., Schmid, C., Pérez, P.: Aggregating local descriptors into a compact image representation. In: 2010 IEEE Conference on Computer Vision and Pattern Recognition (CVPR), pp. 3304–3311. IEEE (2010)
21. Juneja, M., Vedaldi, A., Jawahar, C., Zisserman, A.: Blocks that shout: distinctive parts for scene classification. In: Proceedings of the IEEE Conference on Computer Vision and Pattern Recognition, pp. 923–930 (2013)
22. Kang, K., Wang, X.: Fully convolutional neural networks for crowd segmentation. arXiv preprint arXiv:1411.4464 (2014)

23. Krizhevsky, A., Sutskever, I., Hinton, G.E.: Imagenet classification with deep convolutional neural networks. In: Advances in Neural Information Processing Systems, pp. 1097–1105 (2012)

24. Lazebnik, S., Schmid, C., Ponce, J.: Beyond bags of features: spatial pyramid matching for recognizing natural scene categories. In: 2006 IEEE Conference on Computer Vision and Pattern Recognition (CVPR), vol. 2, pp. 2169–2178 (2006)

25. Lowe, D.G.: Distinctive image features from scale-invariant keypoints. Int. J. Comput. Vis. **60**(2), 91–110 (2004)

26. Mandar, D., Chen, S., Gao, D., Rasiwasia, N., Nuno, V.: Scene classification with semantic fisher vectors. In: Proceedings of the IEEE Conference on Computer Vision and Pattern Recognition, pp. 2974–2983 (2015)

27. Nilsback, M.E., Zisserman, A.: Automated flower classification over a large number of classes. In: Sixth Indian Conference on Computer Vision, Graphics & Image Processing, ICVGIP 2008, pp. 722–729. IEEE (2008)

28. Oliva, A., Torralba, A.: Modeling the shape of the scene: a holistic representation of the spatial envelope. Int. J. Comput. Vis. **42**(3), 145–175 (2001)

29. Quattoni, A., Torralba, A.: Recognizing indoor scenes. In: IEEE Conference on Computer Vision and Pattern Recognition, CVPR 2009, pp. 413–420. IEEE (2009)

30. Simon, M., Rodner, E.: Neural activation constellations: Unsupervised part model discovery with convolutional networks. In: International Conference on Computer Vision (ICCV) (2015)

31. Simonyan, K., Zisserman, A.: Very deep convolutional networks for large-scale image recognition. arXiv preprint arXiv:1409.1556 (2014)

32. Szegedy, C., et al.: Going deeper with convolutions. In: Proceedings of the IEEE Conference on Computer Vision and Pattern Recognition, pp. 1–9 (2015)

33. Uijlings, J.R., Van De Sande, K.E., Gevers, T., Smeulders, A.W.: Selective search for object recognition. Int. J. Comput. Vis. **104**(2), 154–171 (2013)

34. Vogel, J., Schiele, B.: Natural scene retrieval based on a semantic modeling step. In: Enser, P., Kompatsiaris, Y., O'Connor, N.E., Smeaton, A.F., Smeulders, A.W.M. (eds.) CIVR 2004. LNCS, vol. 3115, pp. 207–215. Springer, Heidelberg (2004). https://doi.org/10.1007/978-3-540-27814-6_27

35. Wah, C., Branson, S., Welinder, P., Perona, P., Belongie, S.: The caltech-ucsd birds-200-2011 dataset (2011)

36. Wang, J., Yang, J., Yu, K., Lv, F., Huang, T., Gong, Y.: Locality-constrained linear coding for image classification. In: 2010 IEEE Conference on Computer Vision and Pattern Recognition (CVPR), pp. 3360–3367 (2010)

37. Wang, Z., Feng, J., Yan, S., Xi, H.: Linear distance coding for image classification. IEEE Trans. Image Process. **22**(2), 537–548 (2013)

38. Xiao, J., Hays, J., Ehinger, K.A., Oliva, A., Torralba, A.: Sun database: large-scale scene recognition from abbey to zoo. In: 2010 IEEE Conference on Computer Vision and Pattern Recognition (CVPR), pp. 3485–3492. IEEE (2010)

39. Xiao, T., Xu, Y., Yang, K., Zhang, J., Peng, Y., Zhang, Z.: The application of two-level attention models in deep convolutional neural network for fine-grained image classification. In: 2015 IEEE Conference on Computer Vision and Pattern Recognition (CVPR), pp. 842–850. IEEE (2015)

40. Yang, J., Yu, K., Gong, Y., Huang, T.: Linear spatial pyramid matching using sparse coding for image classification. In: 2009 IEEE Conference on Computer Vision and Pattern Recognition (CVPR), pp. 1794–1801 (2009)

41. Yoo, D., Park, S., Lee, J.Y., Kweon, I.S.: Fisher kernel for deep neural activations. arXiv preprint arXiv:1412.1628 (2014)

42. Yoo, D., Park, S., Lee, J.Y., So Kweon, I.: Multi-scale pyramid pooling for deep convolutional representation. In: Proceedings of the IEEE Conference on Computer Vision and Pattern Recognition Workshops, pp. 71–80 (2015)
43. Zhang, N., Donahue, J., Girshick, R., Darrell, T.: Part-based R-CNNs for fine-grained category detection. In: Fleet, D., Pajdla, T., Schiele, B., Tuytelaars, T. (eds.) ECCV 2014. LNCS, vol. 8689, pp. 834–849. Springer, Cham (2014). https://doi.org/10.1007/978-3-319-10590-1_54
44. Zhang, N., Farrell, R., Iandola, F., Darrell, T.: Deformable part descriptors for fine-grained recognition and attribute prediction. In: Proceedings of the IEEE International Conference on Computer Vision, pp. 729–736 (2013)
45. Zhou, B., Lapedriza, A., Khosla, A., Oliva, A., Torralba, A.: Places: a 10 million image database for scene recognition. IEEE Trans. Pattern Anal. Mach. Intell. **40**, 1452–1464 (2017)
46. Zhou, B., Lapedriza, A., Xiao, J., Torralba, A., Oliva, A.: Learning deep features for scene recognition using places database. In: Advances in Neural Information Processing Systems, pp. 487–495 (2014)

Earth Mover's Distance Between Rooted Labeled Unordered Trees Formulated from Complete Subtrees

Taiga Kawaguchi[1], Takuya Yoshino[2], and Kouichi Hirata[2(✉)]

[1] Graduate School of Computer Science and Systems Engineering,
Kyushu Institute of Technology, Kawazu 680-4, Iizuka 820-8502, Japan
kawaguchi@dumbo.ai.kyutech.ac.jp
[2] Department of Artificial Intelligence, Kyushu Institute of Technology,
Kawazu 680-4, Iizuka 820-8502, Japan
{yoshino,hirata}@dumbo.ai.kyutech.ac.jp

Abstract. In this paper, we introduce earth mover's distances (EMDs, for short) for rooted labeled trees formulated from complete subtrees. First, we formulate the EMDs whose signatures are all of the pairs of a complete subtree and the ratio of its frequency and whose ground distances are either the tractable variations of the tree edit distance, provided from the restricted mappings in the Tai mapping hierarchy, or the complete subtree histogram distance. Then, we show that all of the EMDs are metrics and we can compute them in $O(n^3 \log n)$ time, where n is the maximum number of nodes in given two trees. Finally, we give experimental results of computing EMDs for several data, by comparing the EMDs with their ground distances.

1 Introduction

An *earth mover's distance* (*EMD*, for short) has originally introduced to develop to compare with two images in image retrieval and pattern recognition [10]. The EMD is formulated as the solution of the *transportation problem* between the distributions of features as a pair of a feature and its weight, called *signatures*, in two images, where we call a distance between two signatures a *ground distance*. It is known that the EMD is a metric if so is the ground distance between two signatures and it can be computed in $O(s^3 \log s)$ time, where s is the maximum number of two signatures.

The main purpose of this paper is to formulate a new EMD between *rooted labeled unordered trees* (*trees*, for short), in order to compare tree-structured data such as HTML and XML data for web mining, RNA secondary structures and

This work is partially supported by Grant-in-Aid for Scientific Research 17H00762, 16H02870, 16H01743 and 15K12102 from the Ministry of Education, Culture, Sports, Science and Technology, Japan.

This paper is the improved and the extended version of the paper [8] by partially incorporating with [7].

M. De Marsico et al. (Eds.): ICPRAM 2018, LNCS 11351, pp. 65–88, 2019.
https://doi.org/10.1007/978-3-030-05499-1_4

glycan structures for bioinformatics, phase trees in natural language processing, and so on. Concerned with the EMD for trees, Gollapudi and Panigrahy [4] have extended the EMD to that between two *leaf-labeled trees* whose leaves are assigned to labels but internal nodes are not with the same height.

However, we cannot extend their EMD to be applicable to two standard trees, that is, labels are assigned to all the nodes and having possible different height directly as follows. Their EMD first compares each pair of leaves (that is, the nodes with height 1) and sets the value 1 if both leaves have the same label and 0 otherwise. Then, it computes the distance value of the pair of nodes in the height k, by solving the transportation problem based on the information of the pair of nodes in the height $k - 1$. Hence, in order to apply such a recursion to trees as same as their EMD, the trees are necessary to have the same height and have no internal nodes with labels.

In order to formulate the EMD for not only such leaf-labeled trees but also standard trees, in this paper, we formulate a new EMD between two trees from *complete subtrees* [7,8]. Here, a complete subtree rooted at some node in a tree is a subtree of the tree containing all the descendants of a given root, and the number of all the complete subtrees in a tree is bounded by the number of nodes in the tree.

In order to formulate the EMD between trees, it is necessary to introduce a feature, a signature and a ground distance that is a metric. In this paper, we adopt a complete subtree as a feature, and a pair of a complete subtree and the ratio of its frequencies occurring in a whole tree as a signature. On the other hand, as a ground distance, we adopt the tractable variations of the *edit distance* [12], which is the most famous distance measures between trees.

The edit distance [3,9,12] is formulated as the minimum cost of *edit operations*, consisting of a *substitution*, a *deletion* and an *insertion*, applied to transform from a tree to another tree. Whereas the edit distance is a metric, the problem of computing the edit distance is MAX SNP-hard even if trees are binary [5,18].

Many variations of the edit distance have developed as more structurally sensitive distances [2,6,9,11,15–17,19], as the minimum cost of the restrictions of the Tai mappings. Note that the restrictions of the Tai mapping and then the variations of the edit distance constitute the hierarchy [15]. Almost variations are metrics except an alignment distance [6]. On the other hand, the problem of computing the alignment distance or the bottom-up distance between trees is also MAX SNP-hard [6,14].

Hence, as a ground distance, we adopt the following 5 tractable variations of the edit distance, that is, top-down distance [2,11], an LCA- and root-preserving distance [15], an LCA-preserving distance [19], an accordant distance [9], and an isolated-subtree distance [16,17]. The reason why the variations are tractable is that, when computing them, we can apply the *maximum weight bipartite matching problem* after decomposing trees from the root [14]. In particular, we can regard the minimum weighted bipartite problem as a special case of the transportation problem computing the EMD. In addition, as another ground distance, we also adopt a *complete subtree histogram distance* [1], which is an L_1-distance between histograms for all the complete subtrees.

By combining the signature based on the complete subtrees and the above 6 ground distances, we formulate 6 kinds of EMDs for trees. Then, we show that all the EMDs are always metrics and we can compute all the EMDs in $O(n^3 \log n)$ time, where n is the maximum number of nodes in given two trees.

Finally, we give experimental results to evaluate the EMDs to compare them with their ground distances and investigate the properties of the EMDs. Here, we use real data such as not only N-glycan data from KEGG[1] [7,8] but also all the glycan data from KEGG, PPI corpora[2], CSLOGS[3] and dblp[4].

2   Preliminaries

In this section, we prepare the notions of an earth mover's distance for signatures, trees and an edit distance for trees necessary to discuss later.

2.1   Earth Mover's Distance

We call the set of pairs of a feature p_i and its weight w_i a *signature* and denote it by $P = \{(p_i, w_i) \mid i \in I\}$. For a feature p_i such that $(p_i, w_i) \in P$, we also denote $p_i \in P$ simply.

Let $P = \{(p_i, u_i) \mid i \in I\}$ and $Q = \{(q_j, v_j) \mid j \in J\}$ be signatures. We call a distance between p_i and q_j a *ground distance* and denote it by $gd(p_i, q_j)$. Also we denote the *flow* from p_i to q_j by f_{ij}. When the *cost* of the flow from p_i to q_j is given by $gd(p_i, q_j)f_{ij}$, the overall cost of the flows from P to Q is defined as:

$$\sum_{p_i \in P} \sum_{q_j \in Q} gd(p_i, q_j) f_{ij}.$$

Then, we call the problem of finding the minimum cost flow f_{ij}^* subject to the following constraints a *transportation problem* from P to Q and the minimum cost flow the *optimum flow*:

1. $f_{ij} \geq 0$.
2. $\displaystyle\sum_{p_i \in P} f_{ij} \leq u_i$.
3. $\displaystyle\sum_{q_j \in Q} f_{ij} \leq v_j$.
4. $\displaystyle\sum_{p_i \in P} \sum_{q_j \in Q} f_{ij} = \min\left(\sum_{p_i \in P} u_i, \sum_{q_j \in Q} v_j\right)$.

---

[1] KEGG: Kyoto Encyclopedia of Genes and Genomes, http://www.kegg.jp/.
[2] Protein-protein interaction corpora, http://mars.cs.utu.fi/PPICorpora.
[3] CSLOGS, http://www.cs.rpi.edu/~zaki/www-new/pmwiki.php/Software/Software.
[4] dblp: computer science bibliography http://dblp.uni-trier.de/.

The constraint (1) allows moving "supplies" from P to Q and not vice versa. The constraints (2) and (3) limit the amount of supplies within the weight. The constraint (4) forces to move the maximum amount of supplies possible.

Definition 1 (Earth Mover's Distance [10]). Let P and Q be signatures and f_{ij}^* the optimum flow of the transportation problem from P to Q. Then, we define an *earth mover's distance* (*EMD*, for short) between P and Q as follows.

$$EMD_{gd}(P,Q) = \frac{\sum_{p_i \in P} \sum_{q_j \in Q} gd(p_i, q_j) f_{ij}^*}{\sum_{p_i \in P} \sum_{q_j \in Q} f_{ij}^*} = \frac{\sum_{p_i \in P} \sum_{q_j \in Q} gd(p_i, q_j) f_{ij}^*}{\min \left(\sum_{p_i \in P} u_i, \sum_{q_j \in Q} v_j \right)}.$$

Note that the EMD allows for *partial matches* when the total weight of a signature is different from that of another signature, which is important for image retrieval applications [10]. We can realize the partial match to transport from a signature whose total weight is smaller than a part of another signature. Also the following theorem holds for the EMD.

Theorem 1 (Rubner et al. [10]). *Suppose that two signatures have the same total weight. If a ground distance is a metric, then so is the EMD. Furthermore, we can compute the EMD in $O(s^3 \log s)$ time, where $s = \max\{|P|, |Q|\}$.*

2.2 Trees

A *tree* T is a connected graph (V, E) without cycles, where V is the set of vertices and E is the set of edges. We denote V and E by $V(T)$ and $E(T)$. The *size* of T is $|V|$ and denoted by $|T|$. We sometime denote $v \in V(T)$ by $v \in T$. We denote an empty tree $(\emptyset, \emptyset)$ by $\emptyset$. A *rooted tree* is a tree with one node r chosen as its *root*. We denote the root of a rooted tree T by $r(T)$.

For each node v in a rooted tree with the root r, let $UP_r(v)$ be the unique path from r to v, that is, $UP_r(v) = (\{v_1, \ldots, v_k\}, E)$ such that $v_1 = r$, $v_k = v$ and $(v_i, v_{i+1}) \in E$ for every i ($1 \leq i \leq k - 1$). The *parent* of $v(\neq r)$, which we denote by $par(v)$, is its adjacent node on $UP_r(v)$ and the *ancestors* of $v(\neq r)$ are the nodes on $UP_r(v) - \{v\}$. We denote the set of all ancestors of v by $anc(v)$. We say that u is a *child* of v if v is the parent of u and u is a *descendant* of v if v is an ancestor of u. We use the ancestor orders $<$ and $\leq$, that is, $u < v$ if v is an ancestor of u and $u \leq v$ if $u < v$ or $u = v$.

Let T be a rooted tree (V, E) and v a node in T. A *complete subtree of* T *at* v, denoted by $T[v]$, is a rooted tree $T' = (V', E')$ such that $r(T') = v$, $V' = \{u \in V \mid u \leq v\}$ and $E' = \{(u, w) \in E \mid u, w \in V'\}$. We denote the (multi)set $\{T[v] \mid v \in T\}$ of all the complete subtrees in T by $cs(T)$. For a complete subtree S in T, we denote the frequency of the occurrences of S in T by $f(S, T)$.

Let T be a rooted tree and v a node in T. We call a node having no children a *leaf* and denote the set of all leaves in T by $lv(T)$. We say that w is the *least common ancestor* of u and v in T, denoted by $u \sqcup v$, if $u \leq w$, $v \leq w$ and there exists no $w' \in T$ such that $w' \leq w$, $u \leq w'$ and $v \leq w'$. The *degree* of v, denoted by $d(v)$, is the number of children of v. The *degree* of T, denoted by $d(T)$, is $\max\{d(v) \mid v \in T\}$. The *height* of v, denoted by $h(v)$, is $\max\{|UP_v(w)| \mid w \in lv(T[v])\}$. The *height* of T, denoted by $h(T)$, is $\max\{h(v) \mid v \in T\}$.

For nodes $u, v \in T$, u is *to the left of* v if $pre(u) \leq pre(v)$ for the preorder number *pre* and $post(u) \leq post(v)$ for the postorder number *post*. We say that a rooted tree is *ordered* if a left-to-right order among siblings is given; *unordered* otherwise. We say that a rooted tree is *labeled* if each node is assigned a symbol from a fixed finite alphabet Σ. For a node v, we denote the label of v by $l(v)$, and sometimes identify v with $l(v)$. In this paper, we call a rooted labeled unordered tree a *tree* simply.

2.3   Edit Distance

The *edit operations* [12] of a tree T are defined as follows, see Fig. 1.

1. *Substitution*: Change the label of the node v in T.
2. *Deletion*: Delete a node v in T with parent v', making the children of v become the children of v'. The children are inserted in the place of v as a subset of the children of v'. In particular, if v is the root in T, then the result applying the deletion is a forest consisting of the children of the root.
3. *Insertion*: The complement of deletion. Insert a node v as a child of v' in T making v the parent of a subset of the children of v'.

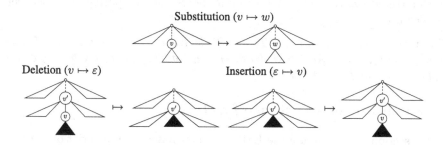

Fig. 1. Edit operations for trees.

Let $\varepsilon \notin \Sigma$ denote a special *blank* symbol and define $\Sigma_\varepsilon = \Sigma \cup \{\varepsilon\}$. Then, we represent each edit operation by $(l_1 \mapsto l_2)$, where $(l_1, l_2) \in (\Sigma_\varepsilon \times \Sigma_\varepsilon - \{(\varepsilon, \varepsilon)\})$. The operation is a substitution if $l_1 \neq \varepsilon$ and $l_2 \neq \varepsilon$, a deletion if $l_2 = \varepsilon$, and an insertion if $l_1 = \varepsilon$. For nodes u and v, we also denote $(l(u) \mapsto l(v))$ by $(u \mapsto v)$. We define a *cost function* $\gamma : (\Sigma_\varepsilon \times \Sigma_\varepsilon - \{(\varepsilon, \varepsilon)\}) \mapsto \mathbf{R}^+$ on pairs of labels. We often constrain a cost function γ to be a *metric*, that is, $\gamma(l_1, l_2) \geq 0$, $\gamma(l_1, l_2) = 0$ iff $l_1 = l_2$, $\gamma(l_1, l_2) = \gamma(l_2, l_1)$ and $\gamma(l_1, l_3) \leq \gamma(l_1, l_2) + \gamma(l_2, l_3)$. In particular, we call the cost function that $\gamma(l_1, l_2) = 1$ if $l_1 \neq l_2$ a *unit cost function*.

Definition 2 (Edit Distance [12]). For a cost function γ, the *cost* of an edit operation $e = l_1 \mapsto l_2$ is given by $\gamma(e) = \gamma(l_1, l_2)$. The *cost* of a sequence $E = e_1, \ldots, e_k$ of edit operations is given by $\gamma(E) = \sum_{i=1}^{k} \gamma(e_i)$. Then, an *edit distance* $\tau_{\mathrm{TAI}}(T_1, T_2)$ between trees T_1 and T_2 is defined as follows:

$$\tau_{\mathrm{TAI}}(T_1, T_2) = \min \left\{ \gamma(E) \,\middle|\, \begin{array}{l} E \text{ is a sequence of edit operations} \\ \text{transforming } T_1 \text{ to } T_2 \end{array} \right\}.$$

Definition 3 (Tai Mapping [12]). Let T_1 and T_2 be trees. We say that a triple (M, T_1, T_2) is an *unordered Tai mapping* (a *mapping*, for short) from T_1 to T_2 if $M \subseteq V(T_1) \times V(T_2)$ and every pair (u_1, v_1) and (u_2, v_2) in M satisfies that (1) $u_1 = u_2$ iff $v_1 = v_2$ (one-to-one condition) and (2) $u_1 \leq u_2$ iff $v_1 \leq v_2$ (ancestor condition). We will use M instead of (M, T_1, T_2) when there is no confusion denote it by $M \in \mathcal{M}_{\mathrm{TAI}}(T_1, T_2)$.

Let M be a mapping from T_1 to T_2. Let I_M and J_M be the sets of nodes in T_1 and T_2 but not in M, that is, $I_M = \{u \in T_1 \mid (u, v) \notin M\}$ and $J_M = \{v \in T_2 \mid (u, v) \notin M\}$. Then, the cost $\gamma(M)$ of M is given as follows.

$$\gamma(M) = \sum_{(u,v) \in M} \gamma(u, v) + \sum_{u \in I_M} \gamma(u, \varepsilon) + \sum_{v \in J_M} \gamma(\varepsilon, v).$$

Theorem 2 (Tai [12]). *The following statement holds.*

$$\tau_{\mathrm{TAI}}(T_1, T_2) = \min\{\gamma(M) \mid M \in \mathcal{M}_{\mathrm{TAI}}(T_1, T_2)\}.$$

Unfortunately, the following theorem holds for computing τ_{TAI} between unordered trees.

Theorem 3 (cf., [1,5,18]). *Let T_1 and T_2 be trees. Then, the problem of computing $\tau_{\mathrm{TAI}}(T_1, T_2)$ is MAX SNP-hard, even if both T_1 and T_2 are binary or the depth of T_1 and T_2 is at most 2.*

3   Earth Mover's Distance for Trees

In this section, we formulate the EMD for trees based on the variations of edit distance. Then, it is necessary to introduce a signature and a ground distance between features.

In this paper, we adopt the following signature $s(T)$ for a tree T.

$$s(T) = \left\{ (S, w) \,\middle|\, S \in cs(T), w = \frac{f(S, T)}{|T|} \right\}.$$

The features of $s(T)$ are complete subtrees of T and the weight of $s(T)$ is the ratio of the occurrences of complete subtrees. Hence, the total weight of s is 1. Since this signature contains T itself, we can transform T to $s(T)$ uniquely.

On the other hand, in order to formulate a ground distance between trees, we introduce tractable variations of the edit distance as the minimum cost of the variations of the Tai mapping.

Definition 4 (Variations of Tai Mapping). Let T_1 and T_2 be trees and $M \in \mathcal{M}_{\mathrm{TAI}}(T_1, T_2)$. We denote $M \setminus \{(r(T_1), r(T_2))\}$ by M^-.

1. We say that M is an *isolated-subtree mapping* [16,17], denoted by $M \in \mathcal{M}_{\mathrm{ILST}}(T_1, T_2)$, if M satisfies the following condition.

$$\forall(u_1, v_1)(u_2, v_2)(u_3, v_3) \in M\Big(u_3 < u_1 \sqcup u_2 \iff v_3 < v_1 \sqcup v_2\Big).$$

2. We say that M is an *accordant mapping* [9], denoted by $M \in \mathcal{M}_{\mathrm{ACC}}(T_1, T_2)$, if M satisfies the following condition.

$$\forall(u_1, v_1)(u_2, v_2)(u_3, v_3) \in M\Big(u_1 \sqcup u_2 = u_1 \sqcup u_3 \iff v_1 \sqcup v_2 = v_1 \sqcup v_3\Big).$$

3. We say that M is an *LCA-preserving mapping* [19], denoted by $M \in \mathcal{M}_{\mathrm{LCA}}(T_1, T_2)$, if M satisfies the following condition.

$$\forall(u_1, v_1)(u_2, v_2) \in M\Big((u_1 \sqcup u_2, v_1 \sqcup v_2) \in M\Big).$$

4. We say that M is an *LCA- and root-preserving mapping* [15], denoted by $M \in \mathcal{M}_{\mathrm{LCART}}(T_1, T_2)$, if $M \in \mathcal{M}_{\mathrm{LCA}}(T_1, T_2)$ and $(r(T_1), r(T_2)) \in M$.
5. We say that M is a *top-down mapping* [2,11], denoted by $M \in \mathcal{M}_{\mathrm{TOP}}(T_1, T_2)$, if M satisfies the following condition.

$$\forall(u, v) \in M^-\Big((par(u), par(v)) \in M\Big).$$

The variations of the Tai mapping provides the following hierarchy [9,15]. Note that the equations does not always hold in general.

$$\mathcal{M}_{\mathrm{TOP}}(T_1, T_2) \subseteq \mathcal{M}_{\mathrm{LCART}}(T_1, T_2) \subseteq \mathcal{M}_{\mathrm{LCA}}(T_1, T_2)$$
$$\subseteq \mathcal{M}_{\mathrm{ACC}}(T_1, T_2) \subseteq \mathcal{M}_{\mathrm{ILST}}(T_1, T_2) \subseteq \mathcal{M}_{\mathrm{TAI}}(T_1, T_2).$$

Then, we can formulate the following variations of the edit distance by using the variations of the Tai mapping.

Definition 5 (Variations of Edit Distance). We define the distance $\tau_A(T_1, T_2)$ as follows, where $A \in \{\mathrm{ILST}, \mathrm{ACC}, \mathrm{LCA}, \mathrm{LCART}, \mathrm{TOP}\}$.

$$\tau_A(T_1, T_2) = \min\{\gamma(M) \mid M \in \mathcal{M}_A(T_1, T_2)\}.$$

We call τ_{ILST} an *isolated-subtree distance* [16,17], τ_{ACC} an *accordant distance* [9], τ_{LCA} an *LCA-preserving distance* [19], τ_{LCART} an *LCA- and root-preserving distance* [15], and τ_{TOP} a *top-down distance* [2,11], respectively.

The above hierarchy of the Tai mapping implies the following inequality for the variations of the edit distance.

$$\tau_{\mathrm{TAI}}(T_1, T_2) \le \tau_{\mathrm{ILST}}(T_1, T_2) \le \tau_{\mathrm{ACC}}(T_1, T_2)$$
$$\le \tau_{\mathrm{LCA}}(T_1, T_2) \le \tau_{\mathrm{LCART}}(T_1, T_2) \le \tau_{\mathrm{TOP}}(T_1, T_2).$$

Furthermore, for all the above variations, the following theorem holds.

Theorem 4 (*cf.*, [14,15,19]). *Let T_1 and T_2 be trees, $n = \max(|T_1|, |T_2|)$ and $d = \min\{d(T_1), d(T_2)\}$. Then, for every $A \in \{\text{ILST}, \text{ACC}, \text{LCA}, \text{LCART}, \text{TOP}\}$, $\tau_A(T_1, T_2)$ is a metric and we can compute $\tau_A(T_1, T_2)$ in $O(n^2 d)$ time.*

In addition, we also introduce a *complete subtree histogram distance* $\tau_{Cs}(T_1, T_2)$ between trees T_1 and T_2 as an L_1-distance between $cs(T_1)$ and $cs(T_2)$, that is:

$$\tau_{Cs}(T_1, T_2) = \sum_{S \in cs(T_1) \cup cs(T_2)} |f(S, T_1) - f(S, T_2)|.$$

Theorem 5 (*cf.*, [1,13,14]). *Let T_1 and T_2 be trees and $n = \max\{|T_1|, |T_2|\}$. Then, $\tau_{Cs}(T_1, T_2)$ is a metric and we can compute $\tau_{Cs}(T_1, T_2)$ in $O(n)$ time.*

Proof. Since we can compute $cs(T)$ in $O(|T|)$ time by traversing nodes in T in postorder with storing all the descendants, the second statement is obvious. □

It is known that a complete subtree histogram distance provides the constant factor of lower bound of $\tau_{\text{TAI}}(T_1, T_2)$ as follows [1].

$$\tau_{\text{TAI}}(T_1, T_2) \leq \tau_{Cs}(T_1, T_2).$$

Note that a mapping corresponding to τ_{Cs} provides a bottom-up *indel* distance, which is an edit distance not allowing substitution, rather than a bottom-up (edit) distance [14]. Then, τ_{Cs} and τ_A for $A \in \{\text{ILST}, \text{ACC}, \text{LCA}, \text{LCART}, \text{TOP}\}$ are incomparable, because so are their mappings [15].

By combining the signatures and the ground distances, we formulate the following 6 kinds of an *EMD for trees*, where $A \in \{\text{ILST}, \text{ACC}, \text{LCA}, \text{LCART}, \text{TOP}, \text{CS}\}$.

Definition 6 (EMD for Trees). Let T_1 and T_2 be trees. Then, we define an *EMD* between T_1 and T_2, denoted by $EMD_A(T_1, T_2)$, as $EMD_{\tau_A}(s(T_1), s(T_2))$ between signatures $s(T_1)$ and $s(T_2)$ for a ground distance τ_A.

Corollary 1. *For trees T_1 and T_2, $EMD_A(T_1, T_2)$ is a metric.*

Proof. Since a ground distance τ_A is a metric by Theorems 4 and 5 and the total weight of signatures is 1, the statement follows from Theorem 1. □

Theorem 6. *Let T_1 and T_2 be trees and $n = \max\{|T_1|, |T_2|\}$. Then, we can compute $EMD_A(T_1, T_2)$ in $O(n^3 \log n)$ time.*

Proof. By using $s(T_1)$, $s(T_2)$ and $\{\tau_A(T_1[u], T_2[v]) \mid (u, v) \in T_1 \times T_2\}$, we can design the following algorithm to compute $EMD_A(T_1, T_2)$.

1. Construct $s(T_1)$ and $s(T_2)$ from T_1 and T_2.
2. Compute $G = \{\tau_A(T_1[u], T_2[v]) \mid (u, v) \in T_1 \times T_2\}$.
3. Compute $EMD_A(T_1, T_2)$ from G.

It is obvious that the running time of Step 1 is $O(n)$. Also since $|s(T_1)| = |s(T_2)| = O(n)$ and by Theorem 1, the running time of Step 3 is $O(n^3 \log n)$.

Consider Step 2. For $A \in \{\text{ILST}, \text{ACC}, \text{LCA}, \text{LCART}, \text{TOP}\}$, since the algorithms of computing $\tau_A(T_1, T_2)$ can store the value of $\tau_A(T_1[u], T_2[v])$ for every $(u, v) \in T_1 \times T_2$ (cf., [15]) and by Theorem 4, we can compute G in $O(n^2 d)$ time, where $d = \min\{d(T_1), d(T_2)\}$. On the other hand, by using $s(T_1)$ and $s(T_2)$ and by Theorem 5, we can compute the value of $\tau_{\text{Cs}}(T_1[u], T_2[v])$ for every $(u, v) \in T_1 \times T_2$ in $O(n)$ time.

Hence, we can compute $EMD_A(T_1, T_2)$ in $O(n^3 \log n)$ time.   □

4 Experimental Results

In this section, for $A \in \{\text{ILST}, \text{ACC}, \text{LCA}, \text{LCART}, \text{TOP}, \text{Cs}\}$, we give experimental results to evaluate EMD_A by comparing with τ_A and investigates the properties of EMD_A. Here, we assume that a cost function is a unit cost function.

In this section, we use real tree data in repositories such as not only N-glycan data (N-glycans) provided from KEGG (see footnote 3) [7,8] but also all of the glycan data (all-glycans) from KEGG, LLL, IEPA and HPRD50 provided from PPI corpora (see footnote 4), CSLOGS (see footnote 5) and dblp (see footnote 6). We use the computer environment that CPU is Intel Xeon E51650 v3 (3.50 GHz), RAM is 1 GB and OS is Ubuntsu Linux 14.04 (64bit).

Table 1 illustrates the details of data, that is, the minimum value a, the maximum value b and the average value c, which we denote by $\langle [a, b]; c \rangle$, of the number of data (#), the average number of nodes (n), the average degree (d) and the average height (h), respectively.

Since either the size of trees or the number of data in CSLOGS and dblp are too huge to compute the distances, we use 1% data in CSLOGS and 0.01% data in dblp selected by the decreasing order of the size of trees to experiment, which is denoted by CSLOGS$^-$ and dblp$^-$, respectively.

Table 1. The details of data.

Data	#	n	d	h
N-glycans	2,142	$\langle [2, 38]; 11.07 \rangle$	$\langle [1, 3]; 2.07 \rangle$	$\langle [1, 12]; 6.20 \rangle$
all-glycans	10,683	$\langle [1, 54]; 6.38 \rangle$	$\langle [0, 5]; 1.65 \rangle$	$\langle [0, 21]; 3.59 \rangle$
LLL	330	$\langle [27, 249]; 111.81 \rangle$	$\langle [3, 13]; 5.94 \rangle$	$\langle [6, 25]; 12.25 \rangle$
IEPA	817	$\langle [23, 281]; 102.26 \rangle$	$\langle [3, 18]; 5.21 \rangle$	$\langle [6, 29]; 12.25 \rangle$
HPRD50	433	$\langle [21, 166]; 84.75 \rangle$	$\langle [2, 12]; 5.42 \rangle$	$\langle [6, 22]; 11.83 \rangle$
CSLOGS	59,691	$\langle [2, 428]; 12.94 \rangle$	$\langle [1.403]; 4.49 \rangle$	$\langle [1, 85]; 3.43 \rangle$
dblp	5,154,530	$\langle [2, 751]; 8.47 \rangle$	$\langle [1, 750]; 7.46 \rangle$	$\langle [1, 4]; 1.01 \rangle$
CSLOGS$^-$	597	$\langle [112, 428]; 169.24 \rangle$	$\langle [3, 403]; 39.96 \rangle$	$\langle [1, 85]; 24.83 \rangle$
dblp$^-$	515	$\langle [55, 751]; 81.37 \rangle$	$\langle [32, 750]; 80.25 \rangle$	$\langle [1, 3]; 1.01 \rangle$

Table 2. The running time to compute the distances (sec.).

τ_A	ILST	ACC	LCA	LCART	TOP	CS
N-glycans	54.33	61.10	48.90	49.24	43.14	3.09
all-glycans	558.24	577.77	486.66	473.86	434.39	63.07
LLL	139.94	155.40	124.06	125.87	109.85	0.81
IEPA	718.97	796.67	635.27	643.08	566.34	5.34
HPRD50	139.07	153.83	122.87	124.37	109.85	1.16
CSLOGS$^-$	2,034.92	1,984.69	1,783.59	1,779.86	1,705.37	3.06
dblp$^-$	1,000.77	1,015.59	984.34	981.91	978.61	0.28
EMD_A	ILST	ACC	LCA	LCART	TOP	CS
N-glycans	76.13	80.53	69.67	70.03	63.85	113.03
all-glycans	715.34	741.81	630.29	634.86	599.97	1,283.83
LLL	495.13	506.86	475.37	475.67	414.16	617.23
IEPA	2,407.80	2,481.11	2,350.63	2,355.90	2,026.08	2,922.25
HPRD50	390.89	404.93	376.51	377.89	314.06	514.00
CSLOGS$^-$	4,467.43	4,571.13	4,359.54	4,386.73	4,646.09	4,165.65
dblp$^-$	1,219.31	1,231.10	1,199.66	1,201.71	1,177.64	248.90

4.1 Running Time

First, we compare the running time to compute EMD_A and τ_A for all data in Table 1. Table 2 illustrates the running time to compute such distances.

Table 3. The ratio (EMD_A/τ_A) of the running time of computing the EMDs (EMD_A) for that of computing the ground distances (τ_A) in Table 2.

EMD_A/τ_A	ILST	ACC	LCA	LCART	TOP	CS
N-glycans	1.40	1.32	1.42	1.42	1.48	36.58
all-glycans	1.28	1.28	1.34	1.34	1.38	20.36
LLL	3.54	3.26	3.83	3.78	3.77	758.22
IEPA	3.35	3.11	3.70	3.68	3.53	547.03
HPRD50	2.81	2.63	3.06	3.04	2.86	443.49
CSLOGS$^-$	2.20	2.30	2.44	2.46	2.72	1,360.43
dblp$^-$	1.22	1.21	1.22	1.22	1.20	873.22

Tables 1 and 2 show that the running time of both EMD_A and τ_A is increasing when the number of nodes is increasing and the ratio of increasing for EMD_A is larger than that for τ_A. On the other hand, the running time is independent from the number of data and, if the number of nodes is similar, then the running time is increasing when the number of data is increasing, by comparing LLL with IEPA.

Note that whereas τ_{Cs} has the shortest running time in τ_A, EMD_{Cs} has the longest running time in EMD_A except CSLOGS$^-$ and dblp$^-$. The reason is as follows. When computing $\tau_A(T_1, T_2)$ for $A \in \{\text{Ilst}, \text{Acc}, \text{Lca}, \text{LcaRt}, \text{Top}\}$, we have designed the algorithm to compute all of the pairs of complete subtrees in T_1 and T_2 (refer to the proof of Theorem 6). On the other hand, we compute $\tau_{Cs}(T_1, T_2)$ and $EMD_{Cs}(T_1, T_2)$ independently, that is, we compute $EMD_{Cs}(T_1, T_2)$ in $O(n^3 \log n)$ time after computing $\tau_{Cs}(T_1, T_2)$ in $O(n)$ time. Furthermore, the average degrees of CSLOGS$^-$ and dblp$^-$ are large, which is one of the reason that the running time of EMD_{Cs} for CSLOGS$^-$ and dblp$^-$ is smaller than that of EMD_A for $A \in \{\text{Ilst}, \text{Acc}, \text{Lca}, \text{LcaRt}, \text{Top}\}$.

Table 3 illustrates the ratio (EMD_A/τ_A) of the running time of computing the EMDs (EMD_A) for that of computing the ground distances (τ_A) in Table 2. Here, we call it the ratio of EMD_A for τ_A simply.

For $A \in \{\text{Ilst}, \text{Acc}, \text{Lca}, \text{LcaRt}, \text{Top}\}$, Table 3 shows that the ratio of EMD_A for τ_A is between 1.2 and 1.5 for N-glycans, all-glycans and dblp$^-$ and between 2.6 and 3.9 for LLL, IEPA, HPRD50 and CSLOGS$^-$. Also, in the variations, Acc has the smallest ratio for N-glycans, all-glycans, LLL, IEPA and HPRD50, Ilst for CSLOGS$^-$ and Top for dblp$^-$. Furthermore, whereas the ratio of EMD_A for τ_A is $O(n \log n/d)$ in theoretical by Theorems 4 and 6, the ratio is at most 4 in experimental. Then, the problems of computing EMDs are efficient for trees when the number of nodes is at most 430 and the height is at most 30 or the number of nodes is at most 760 but the height is at most 2 as dblp$^-$.

On the other hand, Table 3 shows that the ratio of EMD_{Cs} for τ_{Cs} is much larger than those for other variations. Since the average numbers of nodes in N-glycans, all-glycans, LLL, IEPA, HPRD50, CSLOGS$^-$ and dblp$^-$ in Table 1 are 11.07, 6.38, 111.81, 102.26, 84.75, 169.24 and 81.37, respectively, larger average number of nodes gives larger ratio of EMD_{Cs} for τ_{Cs}. Whereas the ratio of EMD_{Cs} for τ_{Cs} is $O(n^2 \log n)$ in theoretical by Theorems 5 and 6, the ratio in experimental is smaller than the theoretical value when substituting the average number of nodes to n.

4.2 Comparing EMDs with Ground Distances

Next, we compare EMD_A with τ_A for $A \in \{\text{Ilst}, \text{Acc}, \text{Lca}, \text{LcaRt}, \text{Top}, \text{Cs}\}$ for all the data.

Figures 2, 3, 4 and 5 illustrates the distributions of ground distances τ_A and EMDs EMD_A for $A \in \{\text{Ilst}, \text{Acc}, \text{Lca}, \text{LcaRt}, \text{Top}, \text{Cs}\}$ for all the data. Here, the x-axis is the value of the distance and the y-axis is the percentage of pairs with the distance pointed by the x-axis. Note that, since the distribution of EMD_{Cs} is different from that of EMD_A for $A \in \{\text{Ilst}, \text{Acc}, \text{Lca}, \text{LcaRt}, \text{Top}\}$, we add the distributions of EMD_A except Cs, which we denote EMD_A^- in these figures.

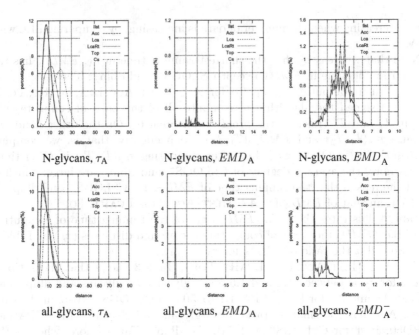

Fig. 2. The distributions of τ_A, EMD_A and EMD_A^- for N-glycans and all-glycans.

Figures 2, 3, 4 and 5 show that, since the values of τ_A and EMD_A for $A \in$ {ILST, ACC, LCA, LCART} are similar and overlapped in graphs, there mainly exist three differences of the values in graphs for $A \in$ {ILST, ACC, LCA, LCART}, TOP and CS except dblp$^-$. For N-glycans and all-glycans in Fig. 2, IEPA in Fig. 3, HPRD50 in Fig. 4 and CSLOGS$^-$ in Fig. 5, τ_A is left to τ_{TOP} and τ_{TOP} is left to τ_{CS}. On the other hand, for LLL in Fig. 3, τ_A is left to τ_{CS} and τ_{CS} is left to τ_{TOP}, and τ_{TOP} has extremely large values.

Furthermore, Fig. 2 shows that the distributions of all of the ground distances are near to normal distributions. For other distributions in Figs. 3, 4 and 5 except dblp$^-$, the number of data is too small not to be near to normal distributions. On the other hand, Fig. 5 shows that almost value of τ_A concentrates near to 0 in the distribution of dblp$^-$.

Figures 6, 7, 8 and 9 illustrate the scatter charts for N-glycans and all-glycans, LLL and IEPA, HPRD50 and CSLOGS$^-$ and dblp$^-$, respectively, between the number of pairs of trees with τ_A pointed at the x-axis and that with EMD_A pointed at the y-axis, where $A \in$ {ILST, ACC, LCA, LCART, TOP, CS}. The diameter and the color represent the number of pairs of trees such that longer diameter and deeper color are larger number.

Figures 6, 7, 8 and 9 show that, for all the data, the scatter charts of A for $A \in$ {ILST, ACC, LCA, LCART} are similar and those and the scatter charts of TOP is slightly different; In particular, for N-glycans, all-glycans, IEPA, HPRD50 and dblp$^-$, the scatter chart of TOP occupies wider regions than the scatter

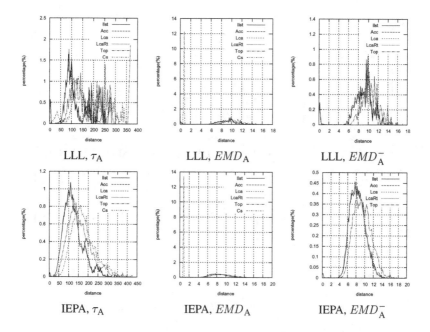

Fig. 3. The distributions of τ_A, EMD_A and EMD_A^- for LLL and IEPA.

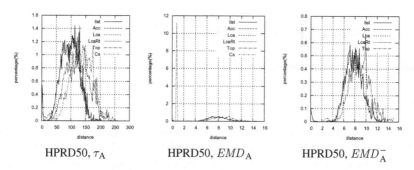

Fig. 4. The distributions of τ_A, EMD_A and EMD_A^- for HPRD50.

charts of other variations. For LLL, the scatter chart of TOP constitutes one cluster, whereas the scatter chars of other variations constitute three clusters.

On the other hand, for the scatter chart of Cs for all the data, while the form is similar as those of other variations, the scale is larger than those of other variations. Hence, the scatter chart of Cs is more expanded than those of other variations. In particular, the scatter charts of Cs for LLL, IEPA and HPRD50 have two clusters.

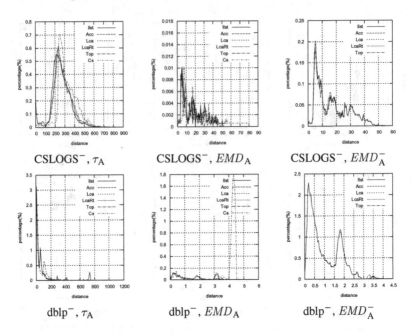

Fig. 5. The distributions of τ_A, EMD_A and EMD_A^- for CSLOGS$^-$ and dblp$^-$.

5   Properties of EMDs for Trees

In this section, we investigate the properties of EMDs for trees.

5.1   Typical Cases

In the following, we point out the typical cases of trees with different values between of τ_A and EMD_A ($A \in \{$ILST, ACC, LCA, LCARST, TOP, CS$\}$) for N-glycans. Here, let u_i be a node in T_1 such that $pre(u_i) = i$ and v_i a node in T_2 such that $pre(v_i) = i$. Also "G*****" denotes the number of N-glycans.

Example 1. Consider trees T_1 and T_2 illustrated in Fig. 10 (left), that is, one tree (T_1) is obtained by deleting leaves to another tree (T_2).

First, we discuss the case when $A \in \{$ILST, ACC, LCA, LCARST, TOP$\}$. In this case, it holds that $\tau_A(T_1, T_2) \le EMD_A(T_1, T_2)$. For the trees T_1 and T_2 in Fig. 10 (left), it holds that $\tau_A(T_1, T_2) = 1$ and $EMD_A(T_1, T_2) = 1.357$ as follows.

For every $1 \le i \le 6$, it holds that $\tau_A(T_1[u_i], T_2[v_i]) = 1$ and $\tau_A(T_1[u_i], T_2[v_7]) = |T_1[u_i]|$. Since the weight of $T_1[u_i]$ (*resp.*, $T_2[v_i]$) is $1/6$ (*resp.*, $1/7$), the optimum flow consists of the 6 flows from $T_1[u_i]$ to $T_2[v_i]$ whose costs are $1/7$ and the 6 flows from $T_1[u_i]$ to $T_2[v_7]$ whose costs are $1/42$. Then, the cost of the optimum flow is $6(1/7) + (6+5+4+3+2+1)/42 = 57/42 = 1.357 = EMD_A(T_1, T_2)$.

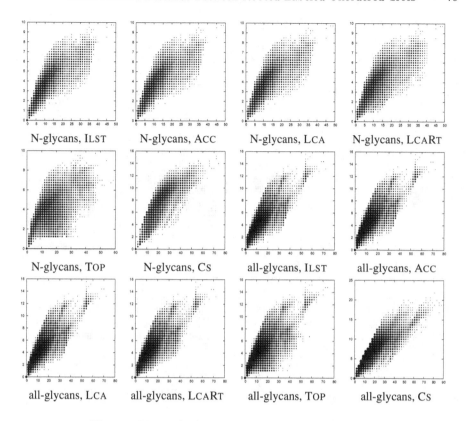

N-glycans, ILST N-glycans, ACC N-glycans, LCA N-glycans, LCART

N-glycans, TOP N-glycans, CS all-glycans, ILST all-glycans, ACC

all-glycans, LCA all-glycans, LCART all-glycans, TOP all-glycans, CS

Fig. 6. The scatter charts for N-glycan and all-glycan.

Hence, whereas the ground distances are not sensitive to inserting leaves, the EMD is necessary to transport the remained weights for every node in one tree to an inserted leave in another tree.

When $A = \text{Cs}$, it holds that $\tau_{\text{Cs}}(T_1, T_2) = 13$ and $EMD_{\text{Cs}}(T_1, T_2) = 7.5$, so it holds that $\tau_{\text{Cs}}(T_1, T_2) > EMD_{\text{Cs}}(T_1, T_2)$, whereas $\tau_A(T_1, T_2) < EMD_A(T_1, T_2)$ for $A \in \{\text{ILST}, \text{ACC}, \text{LCA}, \text{LCART}, \text{TOP}\}$.

Example 2. Consider trees T_1 and T_4 illustrated in Fig. 10 (right), that is, just a label of the root in one tree is different from that in another tree.

First we distance the case for $A \in \{\text{ILST}, \text{ACC}, \text{LCA}, \text{LCART}, \text{TOP}\}$. In this case, it holds that $EMD_A(T_1, T_2) \leq \tau_A(T_1, T_2)$. For the trees T_1 and T_2 in Fig. 10 (right), it holds that $\tau_A(T_1, T_2) = 1$ and $EMD_A(T_1, T_2) = 0.083$ as follows.

The signature containing $r(T_1)$ (*resp.*, $r(T_2)$) is just T_1 (*resp.*, T_2) itself. Since $\tau_A(T_1[u_i], T_2[v_i]) = 0$ for $2 \leq i \leq 12$, the cost of the flow from $T_1[u_i]$ to $T_2[v_i]$ is 0. Since the weight of $T_1[u_i]$ and $T_2[v_i]$ is 1/12 and $\tau_A(T_1[u_1], T_2[v_1]) = 1$, the cost of the optimum flow is $1/12 + 11(0/12) = 0.083 = EMD_A(T_1, T_2)$.

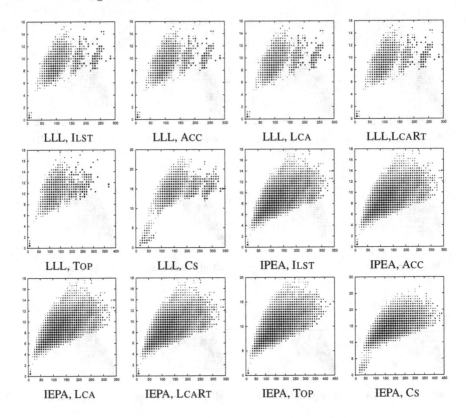

Fig. 7. The scatter charts for LLL and IEPA.

Hence, the difference near to the root is more sensitive to the ground distances rather than the EMDs. Furthermore, in this case, the EMDs is much smaller than the ground distance.

When $A = \text{Cs}$, it holds that $\tau_{\text{Cs}}(T_1, T_2) = 2$ and $EMD_{\text{Cs}}(T_1, T_2) = 0.167$, so it holds that $EMD_{\text{Cs}}(T_1, T_2) < \tau_{\text{Cs}}(T_1, T_2)$, which is same that $EMD_A(T_1, T_2) < \tau_A(T_1, T_2)$ for $A \in \{\text{Ilst}, \text{Acc}, \text{Lca}, \text{LcaRt}, \text{Top}\}$.

Example 3. Consider trees T_i $(1 \leq i \leq 4)$ illustrated in Fig. 11, that is, one tree (T_1 or T_3) is obtained by deleting the root of another tree (T_2 or T_4).

First we distance the case for $A \in \{\text{Ilst}, \text{Acc}, \text{Lca}, \text{LcaRt}, \text{Top}\}$. In this case, it holds that $EMD_{\text{LcaRt}}(T_1, T_2) \leq \tau_{\text{LcaRt}}(T_1, T_2)$ and $EMD_{\text{Top}}(T_3, T_4) \leq \tau_{\text{Top}}(T_3, T_4)$. Here, the values of $A \in \{\text{Ilst}, \text{Acc}\}$ are same as those of Lca. The minimum cost mapping in $\mathcal{M}_A$ is also illustrated in Fig. 11, where the corresponding node is denoted by $\circ$ and the non-corresponding node is denoted by $\bullet$, which implies τ_A.

For the trees T_i in Fig. 11, $\tau_A(T_1, T_2)$, $EMD_A(T_1, T_2)$, $\tau_A(T_3, T_4)$ and $EMD_A(T_3, T_4)$ are as follows.

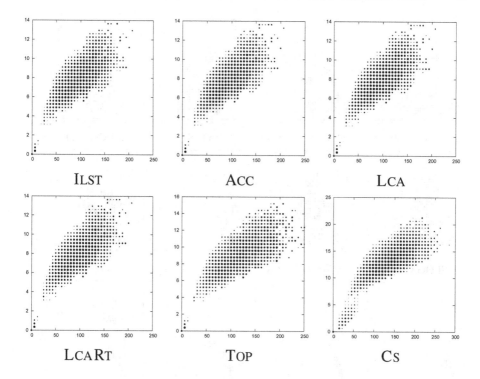

Fig. 8. The scatter charts for HPRD50.

A	$\tau_A(T_1, T_2)$	$EMD_A(T_1, T_2)$	$\tau_A(T_3, T_4)$	$EMD_A(T_3, T_4)$
LCA	2	0.841	2	0.810
LCART	12	0.917	4	0.813
TOP	17	1.512	34	1.092
CS	3	0.931	3	0.852

The reason is that the structural difference near to the root is much sensitive to τ_{LCART} and τ_{TOP}, whose values tend to be large, but the EMDs are not.

The values of Cs have the similar tendency as those of LCA.

Example 4. Consider trees T_1 and T_2 illustrated in Fig. 12, that is, subtrees in one tree (T_1) frequently occur in another tree (T_2).

First we distance the case for $A \in \{\text{ILST}, \text{ACC}, \text{LCA}, \text{LCART}, \text{TOP}\}$. In this case, it holds that $EMD_A(T_1, T_2)$ is much smaller than $\tau_A(T_1, T_2)$. For the trees T_1 and T_2 in Fig. 12, it holds that $\tau_A(T_1, T_2) = 16$ and $EMD_A(T_1, T_2) = 1.63$. Since T_2 is obtained by inserting 16 nodes to T_1, it holds that $\tau_A(T_1, T_2) = 16$.

The weight of $T_1[u]$ (*resp.*, $T_2[v]$) is $1/20$ (*resp.*, $1/36$). Then, $T_1[u_4]$, $T_1[u_{13}]$, $T_2[v_4]$, $T_2[v_{12}]$, $T_2[v_{21}]$ and $T_2[v_{29}]$ are isomorphic and $T_1[u_6]$, $T_1[u_9]$, $T_1[u_{15}]$, $T_1[u_{18}]$, $T_2[v_6]$, $T_2[v_9]$, $T_1[u_{14}]$, $T_1[u_{17}]$, $T_2[v_{23}]$, $T_2[v_{26}]$, $T_1[u_{31}]$ and $T_1[u_{34}]$ are isomorphic, so the weights of $T_1[u_4]$, $T_2[v_4]$, $T_1[u_6]$ and $T_2[u_6]$ as features are

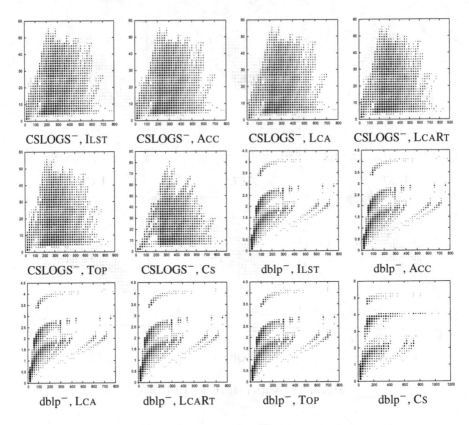

Fig. 9. The scatter charts for CSLOGS⁻ and dblp⁻.

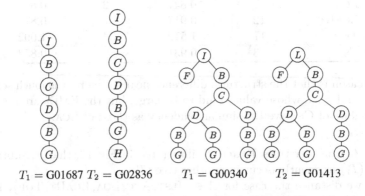

$T_1 = \text{G01687}$ $T_2 = \text{G02836}$ $T_1 = \text{G00340}$ $T_2 = \text{G01413}$

Fig. 10. Trees T_1 and T_2 in Example 1 (left) and Example 2 (right) [8].

2/20, 4/36, 4/20 and 8/36, respectively. Since these weights are preserved in the subtrees of them, the total weight of features consisting of $T_1[u_4]$ and its subtrees in T_1 is $2/20 + 2/20 + 4/20 + 4/20 = 16/20$ and that of $T_2[v_4]$ and its

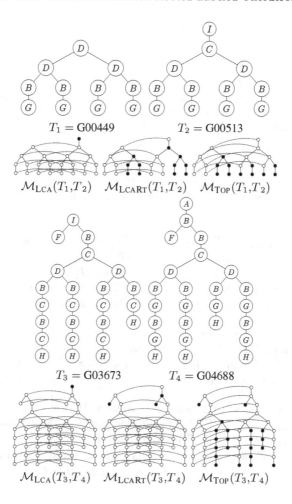

Fig. 11. Trees T_i $(1 \leq i \leq 4)$ in Example 3 [8].

subtrees in T_2 is $4/36 + 4/36 + 8/36 + 8/36 = 32/36$. Hence, the cost of flows in these isomorphic subtrees from T_1 to T_2 is 0, because $\tau_A(T_1[u_4], T_2[v_4]) = 0$, for example. Since these flows move all the weight $16/20$ of $T_1[u_4]$, $T_2[v_4]$ and its subtrees can receive the weight $32/36 - 16/20 = 4/45$.

For the remained features in T_2, the weights of $T_2[v_1]$, $T_2[v_2]$ and $T_2[v_3]$ as features are $1/36$, $1/36$ and $2/36$, respectively. Furthermore, as $T_2[v_4]$ and its subtrees receive the weights, it is necessary to consider the ground distances between $T_1[u_3]$ and $T_2[v_i]$ $(4 \leq i \leq 8)$. The ground distances necessary to compute $EMD_A(T_1, T_2)$ are given as follows.

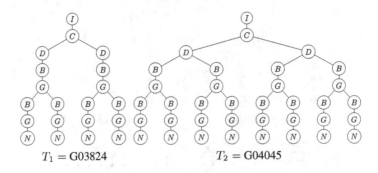

$T_1 = \text{G03824}$ $T_2 = \text{G04045}$

Fig. 12. Trees T_1 and T_2 in Example 4 [8].

$$\tau_A(T_1[u_1], T_2[v_1]) = 16, \ \tau_A(T_1[u_2], T_2[v_2]) = 16,$$
$$\tau_A(T_1[u_3], T_2[v_3]) = 8, \ \tau_A(T_1[u_1], T_2[v_3]) = 3,$$
$$\tau_A(T_1[u_2], T_2[v_3]) = 4, \ \tau_A(T_1[u_3], T_2[v_4]) = 1,$$
$$\tau_A(T_1[u_3], T_2[v_5]) = 2, \ \tau_A(T_1[u_3], T_2[v_6]) = 6,$$
$$\tau_A(T_1[u_3], T_2[v_7]) = 7, \ \tau_A(T_1[u_3], T_2[v_8]) = 8.$$

By computing the optimum flow to receive the weight $4/45 + 4/36 = 1/5$ in T_2, we can obtain $EMD_A(T_1, T_2)$ as $16(1/36) + 16(1/36) + 8(1/90) + 3(1/45) + 4(1/45) + 1(1/90) + 2(1/90) + 6(1/45) + 7(1/45) + 8(1/45) = 49/30 = 1.633$.

When $A = \text{Cs}$, it holds that $\tau_{\text{Cs}}(T_1, T_2) = 24$ and $EMD_{\text{Cs}}(T_1, T_2) = 2.089$.

Example 5. Concerned with Example 1, there exist no pairs of trees T_1 and T_2 such that $\tau_{\text{Cs}}(T_1, T_2) \leq EMD_{\text{Cs}}(T_1, T_2)$ in not only N-glycans but also all-glycans, LLL, IEPA, HPRD50, CSLOGS$^-$ and dblp$^-$.

For N-glycans, T_1 and T_2 in Fig. 13 have the minimum ratio of $\tau_{\text{Cs}}(T_1, T_2)$ to $EMD_{\text{Cs}}(T_1, T_2)$ such that $\tau_{\text{Cs}}(T_1, T_2) = 4$ and $EMD_{\text{Cs}}(T_1, T_2) = 3.00$. On the

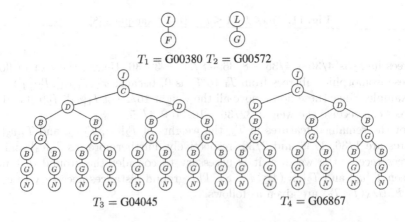

Fig. 13. Trees T_i $(1 \leq i \leq 4)$ in Example 5.

other hand, T_3 and T_4 in Fig. 13 have the maximum ratio such that $\tau_{Cs}(T_3, T_4) = 14$ and $EMD_{Cs}(T_3, T_4) = 1.107$.

In particular, it holds that, for every $A \in \{$ILST, ACC, LCA, LCART, TOP$\}$, $\tau_A(T_1, T_2)$, $EMD_A(T_1, T_2)$, $\tau_A(T_3, T_4)$ or $EMD_A(T_3, T_4)$ is same. The following table summarizes $\tau_{Cs}(T_1, T_2)$, $EMD_{Cs}(T_1, T_2)$, $\tau_A(T_3, T_4)$ and $EMD_A(T_3, T_4)$, which implies that $\tau_A(T_1, T_2) > EMD_A(T_1, T_2)$ and $\tau_A(T_3, T_4) > EMD_A(T_e, T_4)$.

A	$\tau_A(T_1, T_2)$	$EMD_A(T_1, T_2)$	$\tau_A(T_3, T_4)$	$EMD_A(T_3, T_4)$
Cs	4	3	14	1.107
ILST, ACC, LCA, LCART, TOP	2	1.5	8	0.821

5.2 Summary

Finally, we investigate the properties of the EMDs for trees by summarizing the typical cases in Sect. 5.1.

1. Concerned with Example 1, just the case that one tree is obtained by deleting leaves to another tree implies that $\tau_A(T_1, T_2) \leq EMD_A(T_1, T_2)$ for $A \in \{$ILST, ACC, LCA, LCART, TOP$\}$ for N-glycans. Whereas the trees T_1 and T_2 in Example 1 are paths, the statement holds when some internal nodes have some leaves as children. On the other hand, as stated in Example 5, there exists no pair of trees T_1 and T_2 such that $\tau_{Cs}(T_1, T_2) \leq EMD_{Cs}(T_1, T_2)$ in our data.

As another case concerned with Example 1, consider trees T_i $(1 \leq i \leq 6)$ in Fig. 14. Then, for $A \in \{$ILST, ACC, LCA, LCART, TOP$\}$ and Cs, $EMD_A(T_1, T_i)$, $\tau_A(T_1, T_i)$, $EMD_{Cs}(T_1, T_i)$ and $\tau_{Cs}(T_1, T_i)$ are as follows.

T_i	T_2	T_3	T_4	T_5	T_6
$EMD_A(T_1, T_i)$	0.2	0.4	0.6	0.8	1
$\tau_A(T_1, T_i)$	1	1	1	1	1
$EMD_{Cs}(T_1, T_i)$	0.4	1.2	2.4	4	6
$\tau_{Cs}(T_1, T_i)$	2	4	6	8	10

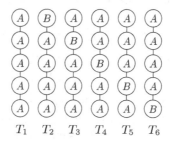

Fig. 14. Trees T_i $(1 \leq i \leq 6)$ in Statement 1 [8].

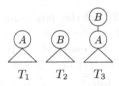

Fig. 15. Trees T_1, T_2 and T_3 in Statement 2 [8].

Then, τ_A is same but τ_{Cs} is increasing when i is increasing. On the other hand, both EMD_A and EMD_{Cs} are increasing when i is increasing. The reason is that the farther node with a different label from the root makes more different signatures.

2. Concerned with Examples 2 and 3, consider complete binary trees T_1 and T_2 with 15 nodes and a tree T_3 adding the root to T_1 illustrated in Fig. 15. Then, for $A \in \{\text{ILST}, \text{TOP}, \text{Cs}\}$, $EMD_A(T_1, T_i)$ and $\tau_A(T_1, T_i)$ are as follows.

T_i	T_2	T_3
$EMD_{\text{ILST}}(T_1, T_i)$	0.067	0.796
$EMD_{\text{TOP}}(T_1, T_i)$	0.067	1.07
$EMD_{\text{Cs}}(T_1, T_i)$	0.133	0.796
$\tau_{\text{ILST}}(T_1, T_i)$	1	1
$\tau_{\text{TOP}}(T_1, T_i)$	1	23
$\tau_{\text{Cs}}(T_1, T_i)$	2	1

Then, it holds that $\tau_{\text{ILST}}(T_1, T_2) = \tau_{\text{ILST}}(T_1, T_3)$, $\tau_{\text{TOP}}(T_1, T_2) < \tau_{\text{TOP}}(T_1, T_3)$ but $\tau_{\text{Cs}}(T_1, T_2) > \tau_{\text{Cs}}(T_1, T_3)$. On the other hand, it holds that $\tau_A(T_1, T_2) < \tau_A(T_1, T_3)$ for $A \in \{\text{ILST}, \text{TOP}, \text{Cs}\}$.

Hence, the difference of both labels and structures near to the root is more sensitive to τ_{TOP} than EMD_{TOP}. On the other hand, for the difference of labels near to the root, EMD_A is much smaller than τ_A. As stated in Examples 2 and 3, there also exists a case that LCATOP is sensitive to the difference of both labels and structures near to the root.

3. Concerned with Example 4, consider a tree T_1 as a path with 10 nodes and trees T_i ($2 \leq i \leq 5$) containing T_1 as subtrees illustrated in Fig. 16. Then, for every $A \in \{\text{ILST}, \text{ACC}, \text{LCA}, \text{LCART}, \text{TOP}, \text{Cs}\}$, $EMD_A(T_1, T_i)$ and $\tau_A(T_1, T_i)$ are as follows.

T_i	T_2	T_3	T_4	T_5
$EMD_A(T_1, T_i)$	0.5	0.738	0.822	0.866
$\tau_A(T_1, T_i)$	1	11	21	31

In this case, whereas the ground distances are necessary to insert new nodes, the EMDs tend to absorb the influence of isomorphic subtrees. Furthermore, τ_{Cs} and EMD_{Cs} have the same values of τ_A and EMD_A for $A \in \{\text{ILST}, \text{ACC}, \text{LCA}, \text{LCART}, \text{TOP}\}$.

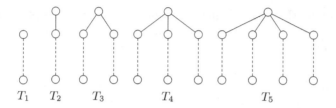

Fig. 16. Trees T_i $(1 \leq i \leq 5)$ in Statement 3.

6 Conclusion

In this paper, we have formulated the earth mover's distances EMD_A based on the variations τ_A of edit distance for $A \in \{\text{ILST}, \text{ACC}, \text{LCA}, \text{LCART}, \text{TOP}, \text{CS}\}$. Then, we have given experimental results to evaluate EMD_A comparing with τ_A. Furthermore, we have investigated the properties of EMD_A.

It is a future work to give experimental results for more large data (with large degrees), for example, AIMed and BioInfer in PPI corpora and larger selected data from CSLOGS and dblp to analyze the theoretical ratios in Sect. 4.1 in experimental. Also it is a future work to formulate EMDs to other tractable variations in Tai mapping hierarchy [15].

Concerned with Table 2 in Sect. 4.1, the running time of EMD_{Cs} is much slower than that of EMD_A for $A \in \{\text{ILST}, \text{ACC}, \text{LCA}, \text{LCART}, \text{TOP}\}$ in experimental, whereas the time complexity is same in theoretical. The reason is that we can compute EMD_A by storing the distance $\tau_A(T_1[u], T_2[v])$ for every pair $(u, v) \in T_1 \times T_2$ when computing $\tau_A(T_1, T_2)$, see the proof of Theorem 6. Then, it is a future work to improve the algorithm to compute EMD_{Cs} by using the same method.

Concerned with Example 1 in Sect. 5.1 and Statement 1 in Sect. 5.2, we have found no trees T_1 and T_2 such that $\tau_A(T_1, T_2) < EMD_A(T_1, T_2)$ for $A \in \{\text{ILST}, \text{ACC}, \text{LCA}, \text{LCART}, \text{TOP}\}$ except the case that T_1 is obtained by deleting leaves to T_2. Then, it is a future work to determine whether or not there exist other cases satisfying that $\tau_A(T_1, T_2) < EMD_A(T_1, T_2)$. On the other hand, concerned with Example 5 in Sect. 5.1 and Statement 1 in Sect. 5.2, there exists no pair of trees T_1 and T_2 such that $\tau_{Cs}(T_1, T_2) \leq EMD_{Cs}(T_1, T_2)$. Then, it is a future work to investigate whether or not this statement always holds in theoretical.

It is a future work to analyze the properties of EMDs in Sect. 5.2 in more detail and investigate how data are appropriate for EMDs. In particular, since it is possible that the number of the signature is too small to formulate EMDs for trees, it is an important future work to investigate appropriate signatures for EMDs for trees.

Concerned with Statement 3 in Sect. 5.2, all the ground distances and all the EMDs (that is, containing Cs) are same when one tree is a path and another is a tree containing the path as subtrees. Hence, it is a future work to investigate whether or not there exist other cases that all the ground distances and all the EMDs are same.

References

1. Akutsu, T., Fukagawa, D., Halldórsson, M.M., Takasu, A., Tanaka, K.: Approximation and parameterized algorithms for common subtrees and edit distance between unordered trees. Theoret. Comput. Sci. **470**, 10–22 (2013)
2. Chawathe, S.S.: Comparing hierarchical data in external memory. In: Proceedings of VLDB 1999, pp. 90–101 (1999)
3. Demaine, E.D., Mozes, S., Rossman, B., Weimann, O.: An optimal decomposition algorithm for tree edit distance. ACM Trans. Algo. **6**(1), 2 (2009)
4. Gollapudi, S., Panigrahy, S.: The power of two min-hashes for similarity search among hierarchical data objects. In: Proceedings of PODS 2008, pp. 211–219 (2008)
5. Hirata, K., Yamamoto, Y., Kuboyama, T.: Improved MAX SNP-hard results for finding an edit distance between unordered trees. In: Giancarlo, R., Manzini, G. (eds.) CPM 2011. LNCS, vol. 6661, pp. 402–415. Springer, Heidelberg (2011). https://doi.org/10.1007/978-3-642-21458-5_34
6. Jiang, T., Wang, L., Zhang, K.: Alignment of trees - an alternative to tree edit. Theoret. Comput. Sci. **143**, 137–148 (1995)
7. Kawaguchi, T., Hirata, K.: On earth mover's distance based on complete subtrees for rooted labeled trees. In: Proceedings of SISA 2017, pp. 225–228 (2017)
8. Kawaguchi, T., Hirata, K.: Earth mover's distance for rooted labeled unordered trees based on Tai mapping hierarchy. In: Proceedings of ICPRAM 2018, pp. 159–168 (2018)
9. Kuboyama, T.: Matching and learning in trees. Ph.D. thesis, University of Tokyo (2007)
10. Rubner, Y., Tomasi, C., Guibas, L.J.: The earth mover's distance as a metric for image retrieval. Int. J. Comput. Vision **40**, 99–121 (2007)
11. Selkow, S.M.: The tree-to-tree editing problem. Inform. Process. Lett. **6**, 184–186 (1977)
12. Tai, K.-C.: The tree-to-tree correction problem. J. ACM **26**, 422–433 (1979)
13. Valiente, G.: An efficient bottom-up distance between trees. In: Proceedings of SPIRE 2001, pp. 212–219 (2001)
14. Yamamoto, Y., Hirata, K., Kuboyama, T.: Tractable and intractable variations of unordered tree edit distance. Int. J. Found. Comput. Sci. **25**, 307–329 (2014)
15. Yoshino, T., Hirata, K.: Tai mapping hierarchy for rooted labeled trees through common subforest. Theory Comput. Syst. **60**, 759–783 (2017)
16. Zhang, K.: Algorithms for the constrained editing distance between ordered labeled trees and related problems. Pattern Recog. **28**, 463–474 (1995)
17. Zhang, K.: A constrained edit distance between unordered labeled trees. Algorithmica **15**, 205–222 (1996)
18. Zhang, K., Jiang, T.: Some MAX SNP-hard results concerning unordered labeled trees. Inform. Process. Lett. **49**, 249–254 (1994)
19. Zhang, K., Wang, J., Shasha, D.: On the editing distance between undirected acyclic graphs. Int. J. Found. Comput. Sci. **7**, 43–58 (1996)

TIMIT and NTIMIT Phone Recognition Using Convolutional Neural Networks

Cornelius Glackin[1]([✉]), Julie Wall[2], Gérard Chollet[1], Nazim Dugan[1], and Nigel Cannings[1]

[1] Intelligent Voice Ltd., London, UK
neil.glackin@intelligentvoice.com
[2] University of East London, London, UK

Abstract. A novel application of convolutional neural networks to phone recognition is presented in this paper. Both the TIMIT and NTIMIT speech corpora have been employed. The phonetic transcriptions of these corpora have been used to label spectrogram segments for training the convolutional neural network. A sliding window extracted fixed sized images from the spectrograms produced for the TIMIT and NTIMIT utterances. These images were assigned to the appropriate phone class by parsing the TIMIT and NTIMIT phone transcriptions. The GoogLeNet convolutional neural network was implemented and trained using stochastic gradient descent with mini batches. Post training, phonetic rescoring was performed to map each phone set to the smaller standard set, i.e. the 61 phone set was mapped to the 39 phone set. Benchmark results of both datasets are presented for comparison to other state-of-the-art approaches. It will be shown that this convolutional neural network approach is particularly well suited to network noise and the distortion of speech data, as demonstrated by the state-of-the-art benchmark results for NTIMIT.

Keywords: Phone recognition · Convolutional Neural Network
TIMIT · NTIMIT

1 Introduction

Automatic Speech Recognition (ASR) typically involves multiple successive layers of hand-crafted feature extraction steps. This compresses the huge amounts of data produced from the raw audio ensuring that the training of the ASR does not take an unreasonably long time. With the adoption of GPGPUs and the so-called Deep Learning trend in recent years, data-driven approaches have overtaken the more traditional ASR pipelines. This means that audio data is automatically processed in its frequency form (e.g. spectrogram) with a Deep Neural Network (DNN), or more appropriately, since speech is temporal, a Recurrent Neural Network (RNN). These networks automate the feature extraction process and can be trained quickly with GPUs. The RNN then converts the spectrogram directly to phonetic symbols or text [1].

Of all the deep-learning technologies, Convolutional Neural Networks (CNNs) arguably demonstrate the most automated feature extraction pipeline. In this paper we have employed CNNs to process the spectrograms as they are well-known for their

© Springer Nature Switzerland AG 2019
M. De Marsico et al. (Eds.): ICPRAM 2018, LNCS 11351, pp. 89–100, 2019.
https://doi.org/10.1007/978-3-030-05499-1_5

state of the art performance for image processing tasks, and this has been adapted for learning the acoustic model component of an ASR system. The acoustic model is responsible for extracting acoustic features from speech and classifying them to symbol classes. The phonetic transcription of both the TIMIT and NTIMIT corpora will be used as the 'ground truths' for training, validation and testing the CNN acoustic model. This consists of spectrograms as input and the phones as class labels. The work presented in this paper builds on previously published work [2] but extends from TIMIT classification to additionally employ the NTIMIT speech corpus to achieve state of the art results. In addition, this paper includes more details regarding the analysis of errors in phone classification.

CNNs are inspired by receptive fields in the mammalian brain and have been typically employed for the classification of static images [3]. Mammalian receptive fields can be found in the V1 processing centres of the cortex responsible for vision and in the cochlear nucleus of the auditory processing areas [4]. They work by transforming the firing of sensory neurons depending on spatial input [5]. Typically, an inhibitory region surrounds the receptive field and suppresses any stimulus which is not captured by the bounds of the receptive field. In this way, receptive fields play a feature extraction role.

Fukushima developed the Neocognitron network inspired by the work of Hubel and Wiesel on receptive fields [6, 7]. The Neocognitron network provided an automated way for implementing feature extraction from images. This approach was advanced by LeCun by incorporating the convolution operations now commonplace in CNNs. It was LeCun that coined the term CNN, the most notable example of which was the LeNet5 architecture which was used to learn the MNIST handwritten character data set [8, 9]. LeNet5 was the first network to use convolutions and subsampling or pooling layers.

Since Ciresan's innovative GPU implementation in 2011 [10], CNNs are now typically trained in parallel with a GPU. Selection of a suitable CNN architecture for classification of any data is dependent upon the amount of available resources and data required to train the networks. The depth of the architecture is positively correlated with the amount of training data required to train them. Additionally, for large network architectures, the number of parameters to be optimized becomes a factor. Perhaps the most efficient architecture to date is the GoogLeNet CNN. It has a relatively complex network structure as compared to AlexNets or VGG networks.

GoogLeNet's main contribution is that it uses Inception modules, within which are convolution kernels extracting features of different sizes. There are 1×1, 3×3, and 5×5 pixel convolutions, typically an odd number so that the kernel can be centred on top of the image pixel in question. 1×1 convolutions are also used to reduce the dimensions of the feature vector, ensuring that the number of parameters to be optimised remains manageable. GoogLeNet's reduced number of parameters was a significant innovation to the field. This is in comparison to its fore-runner AlexNet, which has 60 million parameters to GoogLeNet's 4 million [3]. The pooling layer reduces the number of parameters, but its primary function is to make the network invariant to feature translation. The concatenation layer constructs a feature vector for processing by the next layer. This architecture was used in [2] and was retained here for comparison purposes. Intuitively it is arguably the 'right' depth as far as the volume of images available to train it. The volume of images used to train the networks presented here are slightly larger than the 2011 ImageNet dataset [19] that the GoogLeNet architecture was optimized to tackle.

2 Phone Recognition with TIMIT and NTIMIT

In this work, spectrograms derived from the TIMIT corpus have been used to train a CNN to perform acoustic modelling [11]. The TIMIT corpus, which has an accurate phone transcription, was designed in 1993 as a speech data resource for acoustic phonetic studies. It has been used extensively for the development and evaluation of ASR systems. TIMIT is the most accurately transcribed speech corpus in existence as it contains not only transcriptions of the text but also contains accurate timing of phones. This is impressive given that the average English speaker utters 14–15 phones a second. The corpus contains the broadband recordings of 630 people (438 male/192 female) reading ten phonetically rich sentences of eight major dialects of American English. It includes time-aligned orthographic, phonetic and word transcriptions as well as a 16-bit 16 kHz speech waveform file for each utterance. TIMIT was commissioned by DARPA and worked on by many sites, including Texas Instruments (TI) and Massachusetts Institute of Technology (MIT), hence the corpus' name. Figure 1 features a spectrogram and illustrates the accuracy of the word and phone transcription for one of TIMIT's core training set utterances. A sliding window, shown in grayscale, moves over the 16 kHz (Short-Term Fourier Transform) STFT-based spectrogram. The resulting 256 * 256 pixel spectrogram patches are placed into phone classes according to the TIMIT transcription for training, validation and testing.

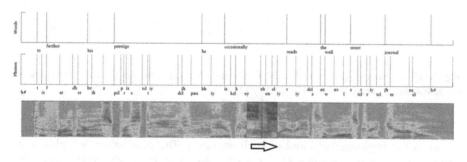

Fig. 1. Presentation of the images for GoogLeNet training [2].

A spectrogram was generated for every 160 samples. For 16 kHz of encoded audio, this corresponds to 10 ms as per the standard resolution required to find all the acoustic features the audio contains. The phone transcriptions are utilised to label each spectrogram according to the phone to which its centre most closely aligns to. Another option would have been to use the centre of the ground truth interval and calculate the Euclidean distance between the centre of the phone interval and the window length. However, this would have made assumptions about where the phone is centred within the interval, requiring an additional computationally expensive step in the labelling of the spectrogram windows.

The previous figure also illustrates the data preparation for the training, validation and testing sets using the sliding window approach. It shows how the phonetic transcription is used to label the 256×256 greyscale spectrogram patches as the sliding window passes over each of the TIMIT utterances. The labelled spectrograms were sorted according to the phone class to which they belong within each of the training, validation and testing sets. In the TIMIT corpus we use the standard core training setup. STFT-type spectrograms were used in particular as they could align the acoustic data and the phonetic symbols with timing that was as accurate as possible. NVIDIA's cuFFT library was used for the FFT component of spectrogram generation [18]. The distribution of the phones that were generated according to the TIMIT ground truth are shown in Fig. 2.

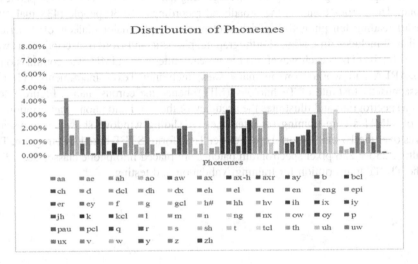

Fig. 2. Distribution of phones within the TIMIT transcription, note that the bars correspond to the alphabetically ordered phones in the key [2].

As evident from Fig. 2, the two largest classes are 's' and 'h#' (silence). Silence occurs at the beginning and end of each utterance. The distribution is highly non-uniform which makes training of the phone classes in the CNN challenging. The training data is the standard TIMIT core set, and the standard test set sub-directories DR1-4 and DR5-8 were used for validation and testing respectively. This partitioning resulted in 1,417,588 spectrogram patches in the training set, as well as 222,789 and 294,101 spectrograms in the validation and testing sets respectively.

2.1 TIMIT GoogLeNet Training and Inferencing

The GoogLeNet acoustic model in this work was trained with Stochastic Gradient Descent (SGD). Prior to the advent of Deep Learning, gradient descent was usually performed using the full batch of training samples in order to adapt the network weights in each training step. However, this approach is not easily parallelizable and thus cannot be implemented efficiently on a GPU. In contrast, SGD computes the gradient of the parameters on a single or few (mini batch) training samples. For larger datasets, such as the one utilised in this work, SGD performs qualitatively as well as batch methods but are faster to train.

We used a stepped learning rate with a 256 sample mini batch size. The resultant training graph is illustrated in Fig. 3. The GoogLeNet architecture produces a phone class prediction at three successive points in the network (loss1, loss2, and loss3). The NVIDIA DIGITS deep learning framework [20] which was employed for this implementation, reports the top-1 and top-5 predictions for each of these loss outputs. loss3 (the last network output) reports the highest accuracy which is 71.65% for classification of the 61 phones. For loss3/top-5 the accuracy is reported as 96.27%, which means that the correct phone was listed in the top five network output classifications over 96% of the time. As mentioned earlier, each spectrogram window contains 4–5 phones on average, and our results show that in the majority of cases these other phones were indeed correctly being reported in the top-5 network outputs.

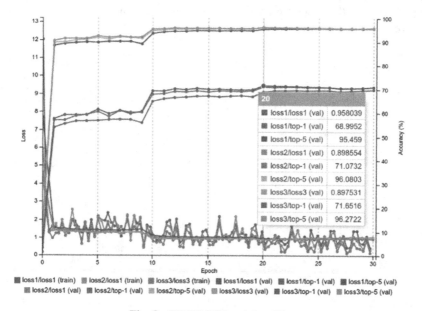

loss1/loss1 (val)	0.958039
loss1/top-1 (val)	68.9952
loss1/top-5 (val)	95.459
loss2/loss1 (val)	0.898554
loss2/top-1 (val)	71.0732
loss2/top-5 (val)	96.0803
loss3/loss3 (val)	0.897531
loss3/top-1 (val)	71.6516
loss3/top-5 (val)	96.2722

loss1/loss1 (train)   loss2/loss1 (train)   loss3/loss3 (train)   loss1/loss1 (val)   loss1/top-1 (val)   loss1/top-5 (val)
loss2/loss1 (val)   loss2/top-1 (val)   loss2/top-5 (val)   loss3/loss3 (val)   loss3/top-1 (val)   loss3/top-5 (val)

Fig. 3. TIMIT SGD training [2].

The network is trained using ∼1.4 M images and uses the validation set (∼223,000 images) to check training progress. A separate set (∼294,000 images) was used to test the system, and the standard test set sub-directories DR1-4 and DR5-8 were used for validation and testing respectively. The validation set is kept separate from the training data and is only used to monitor the progress of the training, and to stop training if overfitting occurs. The highest value of the validation accuracy is used as the final system result and this was achieved at epoch 20, as can be seen in Fig. 3. With this final version of the system, we performed inferencing over the test set, Fig. 4 shows an example prediction from the system for a single sample of unseen test data. The output of the inferencing process contains many duplicates of phones due to the small increments of the sliding window position.

The 256 ms spectrogram windows typically can contain between 4 and 5 phones, with the average speaker uttering approximately 15 phones per second. The pooling layers in the CNN acoustic model provide flexibility in where the feature under question (phones in this case) can be within the 256 * 256 spectrogram image. This is useful for different orientations and scales of images in image classification and is also particularly useful for phone recognition where it is likely there will exist small errors in the training transcription.

During inferencing, the CNN acoustic model makes softmax predictions of all the phone classes for each of the test spectrograms, at three successive output stages of the network (Loss 1 to 3). We conducted some graphical analysis of the output confidences of the phones, colour coding the outputs for easier readability of the results, see Fig. 4. As can be seen from the loss-3 (accuracy), the network makes crisp classifications of usually only a single phone at a time. Given that this is unseen data, and that the comparison with the ground truth is good, we are confident that this network is an effective way to train an acoustic model.

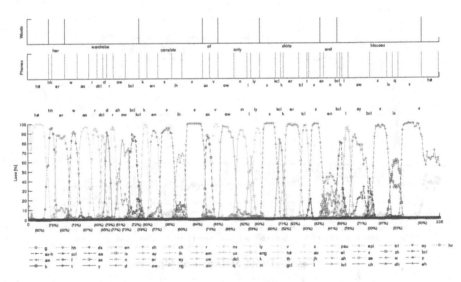

Fig. 4. Softmax network outputs for a test utterance [2].

2.2   Post-processing and Rescoring

Post-processing of the classification output was performed to remove duplicates produced by the fine granularity of the sliding window. It is the convention in the literature when reporting results for the TIMIT corpus to re-score the results for a smaller set of phones [12]. The phoneticians that scored TIMIT used 61 phone symbols. However, many of these phones in TIMIT are not conventionally used by other speech corpora and ASR systems. There are phone symbols called closures e.g. pcl, kcl, tcl, bcl, dcl, and gcl for example, which simply refer to the closing of the mouth before release of closure resulting in the p, k, t, b, d, or g phones being uttered respectively. Most acoustic models map these to the silence symbol 'h#'. Remapping the output of the model inferencing for the unseen testing data to the smaller 39 phone set (See Table 1 for the rescroing mapping), resulted in a significant increase in accuracy from 71.655 (shown in Fig. 3) to 77.44% after rescoring.

This result, while not quite exceeding the 82.3% result reported by Graves [13] with bidirectional LSTMs, or the DNN with stochastic depth [14] which achieved a competitive accuracy of 80.9%, is nevertheless still comparable. The novel approach of Zhang et al. advocates an RNN-CNN hybrid based on MFCC features using conventional MFCC feature extraction with an RNN layer before a deep CNN structure [15]. This hybrid system achieved an impressive 82.67% accuracy. It is not surprising to us that the current state of the art is with a form of CNN [16] with an 83.5% test accuracy. Notably, a team from Microsoft recently presented a fusion system that achieved the state of the art accuracy for the Switchboard corpus. Each of the three ensemble members in the fusion system used some form of CNN architecture, particularly at the feature extraction part of the networks. It is becoming clear that CNNs are demonstrating superiority over RNNs for acoustic modelling.

Table 1. Rescoring mapping of phone symbols (61 to 39 symbols).

aa -> aa	epi -> sil	ow -> ow
ae -> ae	er -> er	oy -> oy
ah -> ah	ey -> ey	p -> p
ao -> aa	f -> sil	pau -> sil
aw -> aw	g -> g	pcl -> sil
ax -> ah	gcl -> sil	q -> sil
ax-h -> ah	h# -> sil	r -> r
axr -> er	hh -> hh	s -> s
ay -> ay	hv -> hh	sh -> sh
b -> b	ih -> ih	t -> t
bcl -> sil	ix -> ih	tcl -> sil
ch -> ch	iy -> iy	th -> th
d -> d	jh -> jh	uh -> uh
dcl -> sil	k -> k	uw -> uw
dh -> dh	kcl -> sil	ux -> uw
dx -> dx	l -> l	v -> v
eh -> eh	m -> m	w -> w
el -> l	n -> n	y -> y
em -> m	ng -> ng	z -> z
en -> n	nx -> n	zh -> sh
eng -> ng		

2.3   NTIMIT Experiments

For commercial speech recognition applications, it is vital to evaluate how ASR performs in the telephone setting. To ensure a fair comparison with the previously published TIMIT speech recognition paper we decided to train the system with NTIMIT [17]. NTIMIT (Network TIMIT) is the result of transmitting the TIMIT database over the telephone network. This results in loss of information in the signal due to the smaller passband of the telephone network, as well as distortions due to the network transmission. To quantify the effects of the network on the original TIMIT data, we calculated the normalized energy of TIMIT and NTIMIT in the [0, 8 kHz] frequency band. Figure 5 shows the absolute value of the normalized amplitudes of the TIMIT (top) and NTIMIT (bottom) corpora.

Both signals are shown here in the [0,8 kHz] range for the 16 kHz sample rate of the original audio as per Nyquist sampling theory. As can be seen from the figure, for TIMIT there is a smooth variation in signal amplitude from the low frequencies to the high frequencies in the entire frequency range. For the NTIMIT signal, it can be seen that after 3400 Hz the gradient of the amplitude stops varying (flatlines) and drops quickly before flatlining again at approximately 6800 Hz (which is likely due to some frequency folding). Consequently, it can be seen that there is very little useful information in the NTIMIT signal above 3400 Hz. Hence for the purposes of the NTIMIT experiments, we first downsampled all of the audio files to 6.5 kHz to ensure that the entire 256 × 256 pixel range resulting spectrogram input to the CNN represents the available speech signal.

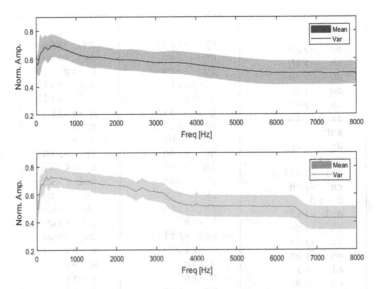

Fig. 5. Spectral profile of TIMIT (top) and NTIMIT (bottom). TIMIT shows a smooth transmission from low to high frequency, whereas NTIMIT has little useful information encoded above 3.3 kHz.

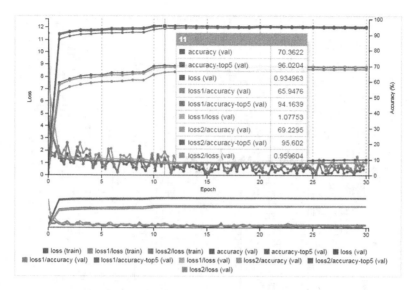

Fig. 6. NTIMIT stochastic gradient descent training.

A GoogLeNet CNN was trained from scratch with spectrograms generated from the downsampled NTIMIT training data. Figure 6 shows progress of the training results in terms of accuracy and loss. The best validation accuracy of 70.36% occurs at epoch 11, the top-5 accuracy achieved is 96.02%. Once again, rescoring was employed to convert the 61 phone set to 39 phones and the accuracy increased to 73.63% as a result.

Figure 7 shows the confusion matrix for the 61 phone set of the NTIMIT test results. The confusion matrix is calculated by accumulating the classification confidences for all 294,000 images in the NTIMIT test set and is informative with regards to the typical pattern of misclassification, in particular with regards to the top-5 performance. It can be quickly understood why the top-5 accuracy is so high, when looking across the rows of the confusion matrix, it can be seen that there are no more than 5 significant classifications for any actual phone. A typical misclassification occurs when comparing the classification of the 'aa' phone as the 'r' sound, although interestingly the results indicate that 'r' is rarely mistaken for 'aa'. The original 61 phone set had many symbols for silence such as 'pau', 'h#', 'sil' and 'q', likely to describe the context of the silence in the original corpus and whether it was a silence at the beginning or end of an utterance ('h#'), a pause ('pau'), etc. However, ultimately an absence of speech regardless of context is easy to mistake when the spectrogram window is small and hence, as can be seen, misclassifications of these various silences are common. Given this, it is understandable why mapping all these silences as well as many of the closure phones, e.g. 'bcl', 'dcl', 'kcl' to 'sil' significantly improves the recognition accuracy. In order to assess how the rescoring improves matters we generated the confusion matrix for the rescored phones in Fig. 8.

Figure 8 shows the confusion matrix for the rescored phone matrices, for the purposes of isolating the remaining phone misclassifications we removed the now amalgamated 'sil' symbols. The main misclassifications can now be seen from the

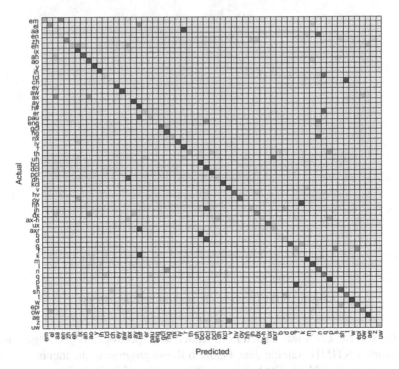

Fig. 7. Confusion matrix of actual against predicted phones.

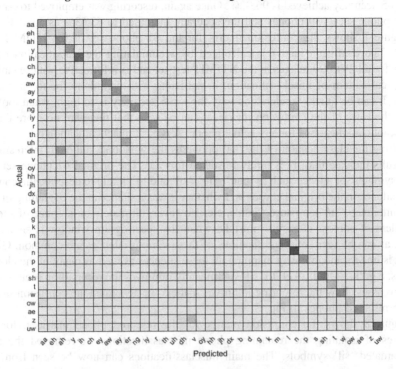

Fig. 8. Confusion matrix of actual against predicted phones for rescored phones.

figure. The 'aa' and 'r' symbols were not rescored and so this misclassification remains, similarly 'ch' and 'sh' are confused by the system. In general, these misclassifications are intuitively understandable and in an ASR system with a good quality language model, many of these typical misclassifications will be overcome by the probability distribution of phonetic sequences inherent in the language model.

3 Conclusions

This paper extended previously published work [2] concerning TIMIT phone classification using CNNs to the case of noisy telephone speech (NTIMIT corpus). Typically, we have found that NTIMIT results in the literature are around 10% less than for TIMIT. However, we have found that we are within 4% of the TIMIT CNN model performance in our tests which suggests that the CNN approach is much more noise robust. To our knowledge the NTIMIT benchmark results reported here are the state of the art.

CNNs are becoming a favoured method for feature extraction of speech data, now more commonly used as input layers to DNN-based ASR systems when used as input to an end to end speech recognition system. The reason for this is that CNNs automate the feature extraction process. We also found in this study, that CNNs seem to be more robust to challenging, noisy and distorted data of the type that is found in the NTIMIT corpus, as compared to other approaches.

References

1. Hannun, A., et al.: Deep speech: scaling up end-to-end speech recognition. arXiv preprint arXiv:1412.5567 (2014)
2. Glackin, C., Wall, J., Chollet, G., Dugan, N., Cannings, N.: Convolutional neural networks for phoneme recognition. In: 7th International Conference on Pattern Recognition Applications and Methods (ICPRAM) (2018)
3. Krizhevsky, A., Sutskever, I., Hinton, G.E.: ImageNet classification with deep convolutional neural networks. In: Advances Neural Information Processing System (NIPS), pp. 1097–1105 (2012)
4. Shamma, S.: On the role of space and time in auditory processing. Trends Cogn. Sci. 5(8), 340–348 (2001)
5. Paulin, M.G.: A method for analysing neural computation using receptive fields in state space. Neural Netw. 11(7), 1219–1228 (1998)
6. Fukushima, K.: Neocognitron: a self-organizing neural network model for a mechanism of pattern recognition unaffected by shift in position. Biol. Cybern. 36(4), 193–202 (1980)
7. Hubel, D.H., Wiesel, T.N.: Receptive fields, binocular interaction and functional architecture in cat's visual cortex. J. Physiol. (London) 160, 106–154 (1962)
8. LeCun, Y., et al.: Handwritten digit recognition with a back-propagation network. In: Advances Neural Information Processing System (NIPS), pp. 396–404 (1990)
9. LeCun, Y.: The MNIST database of handwritten digits (1998). http://yann.lecun.com/exdb/mnist/

10. Ciresan, D.C., Meier, U., Masci, J., Gambardella, L., Schmidhuber, J.: Flexible, high performance convolutional neural networks for image classification. In: International Joint Conference on Artificial Intelligence (IJCAI), vol. 22, no. 1, pp. 1237–1242 (2011)
11. Garofolo, J., et al.: TIMIT Acoustic-Phonetic Continuous Speech Corpus LDC93S1. Linguistic Data Consortium, Web Download, Philadelphia (1993)
12. Lopes, C., Perdigao, F.: Phone recognition on the TIMIT database. In: Speech Technologies/Book 1, pp. 285–302 (2011)
13. Graves, A., Mohamed, A., Hinton, G.: Speech recognition with deep recurrent neural networks. In: IEEE International Conference on Acoustic Speech Signal Process (ICASSP), pp. 6645–6649 (2013)
14. Chen, D., Zhang, W., Xu, X., Xing, X.: Deep networks with stochastic depth for acoustic modelling. In: Signal and Information Processing Association Annual Summit and Conference (APSIPA), pp. 1–4 (2016)
15. Zhang, Z., Sun, Z., Liu, J., Chen, J., Huo, Z., Zhang, X.: Deep recurrent convolutional neural network: improving performance for speech recognition. arXiv 1611.07174 (2016)
16. Tóth, L.: Phone recognition with hierarchical convolutional deep maxout networks. EURASIP J. Audio, Speech, Music Process. 1, 1–13 (2015)
17. Jankowski, C., Kalyanwamy, A., Basson, S., Spitz, J.: NTIMIT: a phonetically balanced, continuous speech, telephone bandwidth speech database. In: IEEE International Conference on Acoustic Speech Signal Processing (ICASSP) (1990)
18. CUDA CUFFT Library: NVIDIA (2007). https://docs.nvidia.com/cuda/cufft/index.html
19. ImageNet Large Scale Visual Recognition Challenge (ILSVRC) (2011). http://image-net.org/challenges/LSVRC/2011/index
20. NVIDIA DIGITS Interactive Deep Learning GPU Training System. https://developer.nvidia.com/digits

An Efficient Hashing Algorithm for NN Problem in HD Spaces

Faraj Alhwarin[✉], Alexander Ferrein, and Ingrid Scholl

Mobile Autonomous Systems & Cognitive Robotics Institute, FH Aachen University of Applied Sciences, Eupener Straße 70, 52066 Aachen, Germany
{alhwarin,ferrein,scholl}@fh-aachen.de

Abstract. Nearest neighbor (NN) search is a fundamental issue in many computer applications, such as multimedia search, computer vision and machine learning. While this problem is trivial in low-dimensional search spaces, it becomes much more difficult in higher dimension because of the phenomenon known as the curse of dimensionality, where the complexity grows exponentially with dimension and the data tends to show strong correlations between dimensions. In this paper, we introduce a new hashing method to efficiently cope with this challenge. The idea is to split the search space into many subspaces based on a number of jointly-independent and uniformly-distributed circular random variables (CRVs) computed from the data points. Our method has been tested on datasets of local SIFT and global GIST features and was compared to locality sensitive hashing (LSH), Spherical Hashing methods (HD and SHD) and the fast library for approximate nearest neighbor (FLANN) matcher by using linear search as a baseline. The experimental results show that our method outperforms all state-of-the-art methods for the GIST features. For SIFT features, the results indicate that our method significantly reduces the search query time while preserving the search quality and outperforms FLANN for datasets of size less than 200 K points.

Keywords: Feature matching · Hash trees · NN search
Curse of dimensionality

1 Introduction

The nearest neighbor (NN) search is one of the most challenging tasks in many computer applications such as machine learning, multimedia databases, computer vision and image processing. In many of these applications, the data points are typically represented as high-dimensional vectors. For example, in computer vision applications, to automatically process and understand an image, it has to be described by one or several of high-dimensional features.

The NN Search problem is defined as follows: Assuming a dataset of points $P \subset \mathbb{R}^d$ are given as $P = \{p_1, p_2, p_3, \ldots, p_n\}$. Then, the problem is to find the

© Springer Nature Switzerland AG 2019
M. De Marsico et al. (Eds.): ICPRAM 2018, LNCS 11351, pp. 101–115, 2019.
https://doi.org/10.1007/978-3-030-05499-1_6

closest point in P to a given query point q using a certain similarity measure such as the hamming or euclidean distance.

$$NN(P,q) = \{p \in P \mid \forall p_i \in P \wedge p_i \neq p : dist(p,q) \leqslant dist(p_i,q)\}.$$

The linear search is the easiest technique to solve this problem, which includes computing all the metric distances from a query point to every point in the dataset for finding the point with the smallest distance (exact NN). However, the query time of linear search grows proportionally to the number and dimensionality of data points. Therefore, this solution is very time-consuming and impractical for large-scale datasets of high-dimensional points.

Much research has been conducted within the last three decades to find an efficient solution for the NN search problem in high-dimensional spaces. The suggested solutions have in common that they organise the dataset content in complex data structures (trees, graphs or hash tables) in such a way that a NN query can be answered without searching the whole dataset. Unfortunately, the time of data saving into and fetching from the data structure keeps growing exponentially with the number of dimensions. The NN search can be accelerated by relaxing the problem by searching for the approximate nearest neighbor (ANN) instead of the exact one. The ANN is defined as any point whose distance to query point is less than $(1+\varepsilon)$ times the distance between the query point and exact NN.

In general, the ANN search algorithms can be typically classified into three major categories: graph-based, tree-based, and hash-based algorithms [1].

The common idea of graph-based algorithms is to build a k-nearest neighbor (kNN) graph in an offline phase. The kNN graph is a network of nodes linked by weighted edges. The nodes represent the data points and edge weights represent the distances between linked points. In the literature, many strategies for kNN graph exploration have been published. In [2], the kNN graph is explored in a best-first order, starting from a few well-separated nodes. In [3], a greedy search is performed, to find the closest node to the query point, staring from a randomly selected starting node.

The tree-based algorithms are based on the recursive partitioning of the search space into sub-spaces.

The most widely used tree-based algorithm is the k-d tree [4,5]. The k-d tree is a k-dimensional binary search tree in which each node represents a partition of the k-dimensional space and the root node represents the whole space. To search for the NN, the coordinates of the query point are used to determine the NN leaf node. Then, the linear search is performed to determine the closest point within the NN leaf. The k-d tree operates successfully in low-dimensional search space, but its performance degrades exponentially with increasing number of dimensions. Various improvements to k-d tree have been suggested [6,7]. Silpa-Anan et al. [8] proposed the use of multiple randomised k-d trees, which are built by selecting the split dimensions randomly from among the dimensions with a high variance. By querying the randomised trees, a single priority queue is used to save answers sorted by their distances to the query point.

In [9], the hierarchical k-means tree is proposed. The hierarchical k-means tree partitions the space hierarchically by using the k-means clustering algorithm.

The k-means algorithm selects k points randomly (called centroids). Then, k initial clusters are created by clustering points according to their distances to the closest centroid. The centroids are iteratively updated to the mean of each cluster, and the clustering is repeated until convergence is achieved.

The most popular hash-based method for ANN search is the locality sensitive hashing (LSH). The LSH was introduced by Indyk et al. [10] for use in binary hamming spaces and later modified by Datar et al. [11] for the use in euclidean spaces. The main idea of LSH is to construct a set of hash functions that project data points from a high-dimensional to one-dimensional space, segmented into intervals of the same length (called buckets). The hash functions are designed, so that the neighbor points in the origin space will be projected to the same hash buckets with high probability [12]. The performance of hash-based algorithms highly relies on the quality of the hash functions used and the tuning of algorithmic parameters. Therefore, many papers dealing with these issues have been proposed (e.g. [13–19]).

Recently, in [20] Heo et al. proposed spherical hashing (SHD) method and compared it with many recent algorithms such as spectral hashing [19], iterative quantization [16], random maximum margin hashing [18] and generalized similarity preserving independent component analysis [17]. They found that their method outperforms all compared state of-the-art algorithms. In [1], Muja and Lowe compared many different algorithms for ANN search using datasets with a wide range of dimensionality. They developed a *Fast Library for ANN search* (FLANN) in high dimensional spaces. FLANN contains a collection of the best algorithms for ANN search and an automatic mechanism for electing the best algorithm and optimal parameters depending on the dataset's content.

In [21], we introduced a hashing method for ANN search in high dimensional spaces. The idea is based on extracting several jointly-independent and uniformly-distributed circular random variables (CRVs) from the data points. These CRVs are then used to index data in hash trees. The CRVs hashing (CRVH) method has been successfully applied to speed up SIFT feature matching. In this paper, we extend our previous work by applying the algorithm on GIST descriptor and by evaluating its performance on large image datasets.

The rest of the paper is organised as follows. In the next Sect. 2, we present CRVH method in more details. In Sect. 3, we show how to apply our method on SIFT and GIST descriptors. In Sect. 4, we evaluate our method on several datasets of SIFT and GIST descriptors (with different sizes) and compare it with the LSH, HD, SHD and FLANN methods. Finally, we conclude the paper in Sect. 5.

2 CRVH Method

In this section, the proposed CRVH method will be theoretically described in details. We start with defining the circular random variables (CRVs).

Definition 1. *Let* $x = (x_0 \ldots x_{n-1})^T$ *be a n-dimensional descriptor vector. We segment it into k segments $s_0, \ldots, s_{k-1}$, each of length l with $k = \lfloor \frac{n}{l} \rfloor$. Each segment s_i is represented by a l-dimensional vector: $s_i = \{x_{i \cdot l}, x_{i \cdot l + 1}, \ldots, x_{i \cdot l + (l-1)}\}$. For each segment, a circular random variable v_i can be defined as the location of the maximum component within the segment. $v_i = \{j \in [0, l-1] \mid x_{i \cdot l + j}$ is the maximum value in $s_i\}$.*

From CRVs, a hash tree with l^k leaves can be constructed. The hash keys $h(x)$ are determined by a polynomial of order $(k-1)$ as follows:

$$h(x) = I = \sum_{i=0}^{k-1} l^i \cdot v_i. \tag{1}$$

Where l denotes the segment length and k the number of segments (the number of CRVs). If the CRVs are jointly-independent and uniformly-distributed, then the data points will be evenly distributed over all the hash tree leaves. In Fig. 1, we show how to extract CRVs from a high-dimensional descriptor. The descriptor is divided into k segments each of length $l = 5$. Form each segment s_i, a CRV v_i is defined as the peak index within the respective segment.

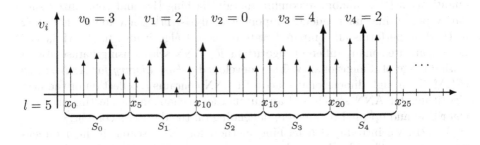

Fig. 1. Extraction of CRVs from a data point represented as d-dimensional vector. In this example, the segment length is chosen equal to 5 [21].

For two neighboring points p_1 and p_2 it is assumed that the CRVs computed from its descriptors tend to be the same and both points are hence hashed into the same hash-tree leaf with high probability

$$Prob\left[h\left(p_1\right) = h\left(p_2\right)\right] > P_1$$
$$Prob\left[h\left(p_1\right) \neq h\left(p_2\right)\right] < P_2$$

with $P_1 \gg P_2$. P_1 is a threshold of the probability that two true neighbors are hashed to the same leaf and P_2 is a threshold of the probability that two true neighbors are hashed to different leaves. For the sake of simplicity, we explain this assumption in two-dimensional search space as shown in Fig. 2. In 2D space,

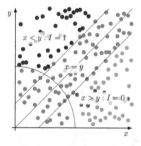

Fig. 2. 2D classification using one CRV [21]. (Color figure online)

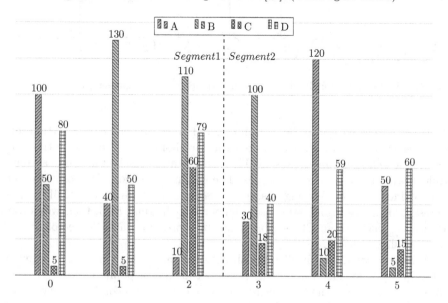

Fig. 3. Vectors of 6D points, A, B, C and D.

the vector can be considered as one segment of length $l = 2$ and hence only one circular binary random variable can be constructed v. If the abscissa of a vector is larger than the ordinate $(x > y)$ we get $v = 0$ (red dots in Fig. 2), otherwise $y > x$ yields $v = 1$ (blue dots in the figure). For boundary points, which whose segments have no dominant maximum (grey dots in Fig. 2), we get a boundary problem. In this case, boundary problem can be avoided by adding boundary points to both hash leaves. In high-dimensional spaces, the boundary problem can be solved by considering not only the maximum indices that define the CRVs, but also the second maximum indices if the second to first maximum ratio is greater than a certain threshold T.

$$T = \frac{max_2}{max_1} \tag{2}$$

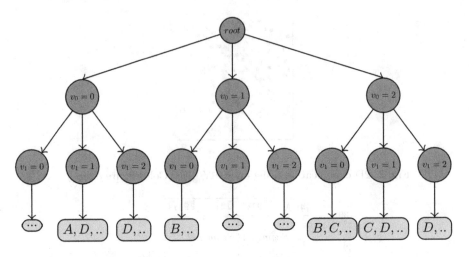

Fig. 4. Hash tree for 6D dataset with two CRVs of segment length $= 3$ [21].

where max_1 and max_2 are the first and second maximum values in a certain segment, respectively. The adjustment of ratio threshold is used to make a trade-off between search speed and precision. This consideration can be taken into account while storing or/and querying stages. The dealing with boundary problem can be explained by the following example: Assuming that we have 4 of 6D points represented in Fig. 3 and we chose segment length equal to 3, then we obtain two CRVs. Using these CRVs, points can be stored into a hash tree of $3^2 = 9$ leaves as shown in Fig. 4.

While querying, we can distinguish between three cases: first, each segment of the query point has a dominant peak as the case of point A (see Fig. 3), in this case, only one of the 9 leaves has to be searched. In second case, one of the segments has no dominant peak (as the case of points B or C). In this case, we have to consider the maximum and the second maximum and hence two leaves have to be searched. The last case is if both segments have no dominant peaks (as in the case of point D). In this case, 4 leaves out of 9 have to be explored. For illustration purposes, Fig. 3 shows the maximum, second maximum and their indices of two segments for 4 example points, and Fig. 4 shows how they are stored (or queried) in the hash tree, when the ratio threshold T is set to 0.5.

Using the above hash function, the dataset points will be evenly distributed over all the hash leaves, if two conditions are met, firstly the used CRVs are uniformly-distributed, and secondly, the CRVs are jointly-independent.

To verify if the CRVs meet the uniformly-distributedness condition, their probability density functions (PDFs) are estimated from a large set of data points. The PDFs are computed by constructing histograms of the CRVs ranging between $[0, l - 1]$. Once the PDFs are estimated, the χ^2 test is used to quantitatively evaluate the goodness fit to the uniform distribution. The value of the test statistic is defined as:

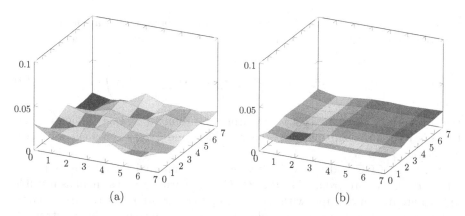

(a)   (b)

Fig. 5. Comparison between (a) joint pdf of two uncorrelated CRVs and (b) the product of their individual pdfs.

$$\chi^2 = \sum_{i=1}^{n} \frac{(O_i - E)^2}{E} \tag{3}$$

where O_i is the observed pdf value, E is the expected value from the uniform distribution and n is the number of possible values the CRVs can take. Mathematically, it is known that, a set of random variables $(V_1, V_2, \ldots, V_n)$ are jointly-independent, if the joint probability density function is equal to the product of their individual pdfs.

$$pdf(V_1, V_2, \ldots, V_n) = \prod_{i=1}^{n} pdf(V_i) \tag{4}$$

To verify whether the CRVs meet the jointly-independent condition, the circular correlation coefficient (CCC) is firstly used to filter out correlated CRVs. The jointly-independent condition is then examined for the remaining uncorrelated CRVs by comparing the joint probability density function with the product of the individual probability functions. In our case, we experimentally find that CCC is sufficient to determine whether the CRVs are jointly-independent or not. In Fig. 5, the joint PDF and the product of PDFs of two uncorrelated CRVs (extracted from one million 960-D GIST descriptors) are shown. As seen by comparing the Fig. 5(a) with Fig. 5(b), for two uncorrelated CRVs, the joint PDF tends to be equal to the individual PDFs'product.

The CCC proposed by Fisher and Lee [22] is defined as:

$$\rho(\alpha, \beta) = \frac{\sum_{i=0}^{n} \sin(\alpha - \bar{\alpha}) \cdot \sin(\beta - \bar{\beta})}{\sqrt{\sum_{i=0}^{n} \sin(\alpha - \bar{\alpha})^2 \cdot \sum_{i=0}^{n} \sin(\beta - \bar{\beta})^2}} \tag{5}$$

where $\rho(\alpha, \beta)$ is the CCC, n is the number of data points, α, β are two circular variables and $\bar{\alpha}$, $\bar{\beta}$ are their respective circular mean values defined as:

$$mean\,(\alpha) = \bar{\alpha} = \arctan\left(\sum_{i=0}^{n} \sin(\alpha), \sum_{i=0}^{n} cos\,(\alpha)\right). \tag{6}$$

the absolute of $\rho(\alpha, \beta)$ takes values from the interval $[0, 1]$. 0 indicates that there is no relationship between the variables, and 1 represents the strongest association possible.

Once the CCC and the χ^2 test values are computed for all the extracted CRVs, the CRVs are grouped so that CCC and χ^2 values are as small as possible in each group. An example with CCCs computed for SIFT descriptors is shown in Fig. 6. For each group of uncorrelated CRVs, a hash tree can be construct.

The above conditions ensure that the data are well-balancedly distributed over all the hash tree leaves. Therefore, the speed-up factor gained by our method compared to linear search can be theoretically determined as follows:

$$(\frac{l}{2})^k \le SF \le l^k \tag{7}$$

where l is CRV length and k is the number of CRVs that meet the conditions mentioned above.

In general, our method consists of three main steps. In the first step we extract the CRVs from the descriptor. To this end, we study the characteristics of the descriptor statistically to determine how its components are distributed and dependent on each other. Based on these characteristics we divide the descriptor into a set of equal length segments. For each segment, a CRV is defined as the location of the maximum within this segment. We calculate then the probability density functions and the joint dependency between the CRVs, and we group them into groups so that they are uniformly-distributed and jointly-independent as much as possible. The goal behind that is to spread the points over hash leaves uniformly and hence to maximize the speed-up factor defined in Eq. 7 regardless of how the points are distributed in their origin space.

In the second step we store the dataset points into hash trees. For this goal, we calculate the CRVs from the segments specified in the first step. From CRVs of each point, a hash key is computed as defined in Eq. 1. The hash key is then used to specify the hash leaves where to store this point. This step is done once for each dataset.

The last, third step is the same as the second one, but it is run on-line and applied to each query point in order to determine the hash leaves, where the candidate neighbors of the query point can be expected.

In the next section, we will describe how to apply this method on local SIFT and global GIST descriptors, which are mostly used in computer vision and image processing applications.

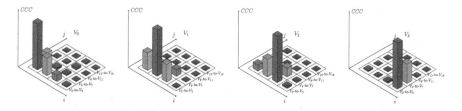

Fig. 6. Plot of circular correlation coefficients for V_0 to V_3 between each two CRVs for SIFT descriptors (out of 16) [21].

3   CRVH Method for SIFT and GIST Descriptors

To apply CRVH method on SIFT descriptors, we choose the CRV segment length $l = 8$. Then, from the 128-dimensional vector, 16 different CRVs can be obtained. A subset of these CRVs have to be selected, so that the CRVs meet the jointly-independent and uniformly-distribution conditions. For this goal, we have statistically analysed a dataset of the SIFT descriptors. Statistically, we found that the SIFT descriptor has a special signature, so that some components are always larger than some others [21]. The signature of SIFT descriptors is represented in Fig. 7 by the mean value of each component. For example, the 41th, 49th, 73th and the 81th components are always significantly larger than their neighbors.

The signature of SIFT descriptors influences the distribution of proposed CRVs. In order to remove this effect, SIFT descriptors are weighted before computing of CRVs. The weight vector is defined as the inverse of individual elements of the signature vector. $S = [s_1, s_2, \ldots, s_n] \Rightarrow W = [s_1^{-1}, s_2^{-1}, \ldots, s_n^{-1}]$ Fig. 8 shows the PDF functions of the CRVs before and after removing the descriptor signature effect. As shown in Fig. 8(b), after removing signature effect, all CRVs meet the uniformly-distributedness condition.

To study the dependence between the CRVs, the circular correlation coefficient (CCC) between each two CRVs is calculated. The CCCs between CRVs are explained in Fig. 6. Figure 6 shows that neighboring CRVs in the descriptor are highly-correlated, whereas there are no or only very weak correlations between non-contiguous CRVs. We omitted the other 12 diagrams showing the correspondences for the other CRVs. From the 16 CRVs we get two sub-groups of jointly-independent and uniformly-distributed CRVs:

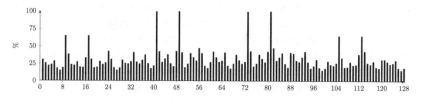

Fig. 7. The signature of a SIFT descriptor; the normalised mean values of the SIFT descriptor components were computed from a dataset of 100 K descriptors.

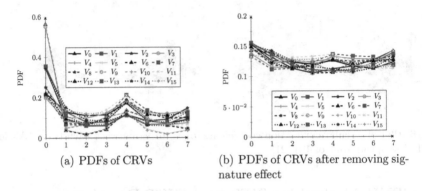

(a) PDFs of CRVs

(b) PDFs of CRVs after removing signature effect

Fig. 8. The probability density functions of the CRVs (extracted from 100k SIFT features) before and after removing signature influence.

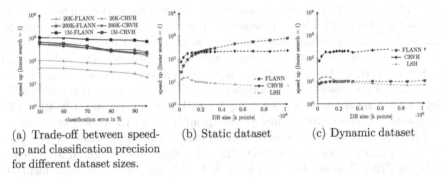

(a) Trade-off between speed-up and classification precision for different dataset sizes.

(b) Static dataset

(c) Dynamic dataset

Fig. 9. Speed-up and precision comparison between CRVH and FLANN; baseline is linear search for static and dynamic datasets.

$g_1 = \{V_0, V_2, V_5, V_7, V_8, V_{10}, V_{13}, V_{15}\}$
and $g_2 = \{V_1, V_3, V_4, V_6, V_9, V_{11}, V_{12}, V_{14}\}$.
From these two sub-groups, two hash trees can be constructed.

With a segment length of 8, 120 or 48 different CRVs can be obtained from 960-D or 384-D GIST descriptors respectively. These CRVs are classified into several sub-groups, so that the CRVs of each group have to meet the jointly-independent and the uniformly-distribution conditions. Similar to SIFT, we statistically found that a GIST descriptor has also a special signature, so that some components are always larger than some others. Before computing CRVs from the GIST descriptor, this signature is neutralised by weighting the descriptor with a constant weight vector. The weight vector is computed as the case of SIFT feature by inverting the signature components. Experimentally, we found that, CRVs of 960-D GIST descriptor can be classified into 14 sub-groups, so that each sub-group contains 6 jointly-independent and uniformly-distributed CRVs. Hence, from these sub-groups, 14 hash trees can be constructed. Similarly, for 384-D GIST descriptor, 6 hash trees can be constructed.

4   Empirical Evaluation

In this section, we analyse the performance of our method on two kinds of descriptors: local SIFT and Global GIST descriptors. We further compare the method with FLANN, LSH, DH and SHD on different image datasets. In the following experiments, we use the OpenCV implementation of FLANN and the C++ implementation of LSH, HD and SHD available in [20]. All experiments were carried out on a Linux machine with an Intel(R) Core(TM) i7-4770S CPU 3.10 GHz and 32 GB RAM.

For SIFT descriptors, the performance of our method is compared with the state-of-the-art NN matcher FLANN and LSH. The experiments are carried out using the *Oxford Buildings Dataset* of real-word images [23]. The *Oxford Buildings Dataset* dataset consists of about 5000 images. Among them there are several pairs that show the same scene from different viewpoints. 10 images of that pairs are taken out of the dataset and used as the query set.

We compare our method with FLANN and LSH in terms of both, speed-up over the linear search (baseline) and the percentage of correctly sought neighbors (precision). To evaluate the performance of our method, two experiments were conducted. The first experiment was carried out with different dataset sizes (20 K, 200 K, 1M) by varying precision parameters. We measured the trade-off between the speed-up and the precision. For the FLANN matcher, the precision was adjusted by varying appropriate FLANN parameters (number of trees and checks), whereas for the CRVH method, the precision was changed by varying the ratio threshold. The obtained results are shown in Fig. 9(a). As can be seen from the figure, our method outperforms the FLANN matcher for datasets with a size of less than 200 K descriptors.

In the second experiment, the performance is compared against FLANN and LSH for two different settings, a static and a dynamic dataset, respectively. In the static setting, the image dataset remains unchanged, while in the dynamic one, the dataset needs to be updated on-line by adding or deleting images. In this experiment, we keep the precision level at 90% and vary the size of dataset. Figure 9 shows the obtained results for both dataset settings. Figure 9(c) shows that in the case of a dynamic datasets, the CRV method outperforms both the LSH and FLANN methods for all dataset sizes significantly. It reaches speed-up factor of 20 over FLANN for dataset sizes up 100 K descriptors. The reason of this outcome can be explained by the FLANN method constructs a specific nearest neighbor search index for a specific dataset; when the dataset is updated by adding or removing some data, the search index has to be updated as well, otherwise the search speed decrease. Conversely, the CRVH method works independently from the dataset contents and its performance is not influenced by adding or removing data points.

For the GIST descriptor, the performance of our method is compared with LSH and Spherical Hashing methods (HD and SHD). The experiments are carried out using the Tiny [24] dataset. The Tiny dataset consists of 80 million 32×32 color images. Each image is described by one GIST descriptor of dimension 960 or 384. We evaluate our method on three different-sized sub-sets of

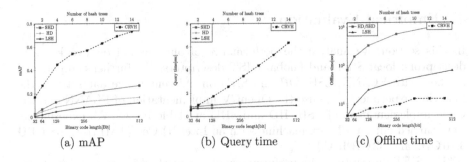

(a) mAP   (b) Query time   (c) Offline time

Fig. 10. Results on 50 k 960-D GIST descriptors.

the Tiny dataset (50K, 1M and 10M images). In each experiment, 100 GIST descriptors are randomly selected and used as query set. The remaining descriptors are used as the dataset. To obtain statistically meaningful results, for 50 K and 1M datasets, the experiments were repeated 5 times with varying the data- and query sets. For 10M, the experiment was repeated only two times because of the time required for computing ground truth. The performance is measured by mean average precision (mAP), query time and offline time. The precision is defined as the fraction of true points among the top Knn retrieved points. The ground truth is determined by the Knn nearest neighbors that are retrieved by the linear search based on Euclidean distance of GIST descriptors. For the LSH, HD, and SHD hashing methods, the performance was measured across code binary lengths ranging from 32 to 512 bits, whereas for CRVH, the performance was measured across all possible number of hash trees (14 trees for 960D and 6 trees for 384D GIST descriptors). For all experiments, we set $Knn = 100$, and we set the ratio threshold T of CRVH to $T = 0.5$. The offline time of LSH is the time required to compute binary code, while for HD/SHD it is the time of binary code computation plus the spherical hashing learning time. For our method, offline time is the time required to construct hash trees.

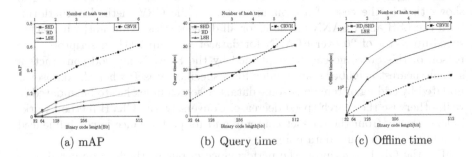

(a) mAP   (b) Query time   (c) Offline time

Fig. 11. Results on 1 million 385-D GIST.

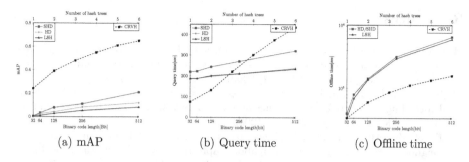

(a) mAP   (b) Query time   (c) Offline time

Fig. 12. Results on 10 million 384-D GIST descriptors.

Figure 10 shows the results on 50 K 960D GIST descriptors. Using two hash trees, our method reaches a mAP (Fig. 10(a)) better than all state-of-the-art methods with a similar query time (see Fig. 10(b)). When increasing the number of threes, our method can reaches $mAP > 0.5$ with a query time less than 6 sec. Regarding offline time, Fig. 10(c) shows that our method extremely outperforms the other methods. For example, for a binary code length of 512, HD and SHD needed about 1200 s, while constructing 14 hash tree only needs about 15 s.

Figures 11 and 12 show the results on 1M and 10M of 384D GIST descriptors. For 3 hash trees, our method reaches $mAP = 0.45$ at query time about 15 s (Fig. 11(a)), while the best state-of-the-art method (SHD) reaches $mAP = 0.22$ at a query time about 30 s. Figures 11 and 12 also demonstrate the scalability of our methods. It is shown that, on both dataset sizes (1M and 10M), our method provides similar mAP, while the query and offline time increases linearly.

5 Conclusions

In this paper, we presented a hashing method for NN search in high-dimensional spaces. Our method bases on extracting a set of CRVs from data points. The data vector is divided into several segments. For each segment, a CRV is defined as the relative position of the peak in that segment. The CRVs are grouped together in such a way that in each group, they are all jointly-independent and uniformly-distributed. The CRVs are exploited to store data points evenly in one or several hash trees. In the query phase, the CRVs of query point are used to determine the hash leaves, where candidate neighbors can be found. The proposed method was tested on datasets of SIFT and GIST descriptors and compared with LSH, HD, SHD and FLANN. The presented experimental results show that, our proposed method outperforms all compared state-of-art methods.

References

1. Muja, M., Lowe, D.: Fast approximate nearest neighbors with automatic algorithm configuration. In: Proceedings of the International Conference on Computer Vision Theory and Applications (VISAPP 2009), pp. 331–340 (2009)
2. Sebastian, B., Kimia, B.B.: Metric-based shape retrieval in large databases. In: Proceedings of the IEEE Conference on Computer Vision and Pattern Recognition (CVPR 2002), vol. 3, pp. 291–296 (2002)
3. Hajebi, K., Abbasi-Yadkori, Y., Shahbazi, H., Zhang, H.: Fast approximate nearest-neighbor search with k-nearest neighbor graph. In: Proceedings of the 22nd International Joint Conference on Artificial Intelligence (IJCAI 2011), pp. 1312–1317 (2011)
4. Bentley, L.: Multidimensional binary search trees used for associative searching. Commun. ACM (CACM) 18(9), 509–517 (1975)
5. Friedman, J.H., Bentley, J.L., Finkel, R.A.: An algorithm for finding best matches in logarithmic expected time. ACM Trans. Math. Softw. 3(3), 209–226 (1977)
6. Beis, J.S., Lowe, D.G.: Shape indexing using approximate nearest-neighbour search in high-dimensional spaces. In: Proceedings of the IEEE Conference on Computer Vision and Pattern Recognition (CVPR 1997), pp. 1000–1006 (1997)
7. Arya, S., Mount, D.M., Netanyahu, N.S., Silverman, R., Wu, A.Y.: An optimal algorithm for approximate nearest neighbor searching in fixed dimensions. J. ACM 45(6), 891–923 (1998)
8. Silpa-Anan, C., Hartley, R.: Optimised KD-trees for fast image descriptor matching. In: Proceedings of the IEEE Conference on Computer Vision and Pattern Recognition (CVPR 2008), pp. 1–8 (2008)
9. Fukunaga, K., Narendra, P.M.: A branch and bound algorithm for computing k-nearest neighbors. IEEE Trans. Comput. (TC) C−24(7), 750–753 (1975)
10. Indyk, P., Motwani, R.: Approximate nearest neighbors: towards removing the curse of dimensionality. In: Proceedings of the Symposium on Computational Geometry (SoCG 1998), pp. 604–613 (1998)
11. Datar, M., Indyk, P., Immorlica, N., Mirrokni, V.S.: Locality-sensitive hashing scheme based on p-stable distributions. In: Proceedings of the Thirtieth Annual ACM Symposium on Theory of computing (STOC 2004), pp. 253–262 (2004)
12. Andoni, A., Indyk, P.: Near-optimal hashing algorithms for approximate nearest neighbor in high dimensions. Commun. ACM 51(1), 117–122 (2008)
13. Kulis, B., Grauman, K.: Kernelized locality-sensitive hashing for scalable image search. In: Proceedings of the IEEE 12th International Conference on Computer Vision (ICCV 2009), pp. 2130–2137 (2009)
14. Wang, J., Kumar, S., Chang, S.F.: Semi-supervised hashing for scalable image retrieval. In: Proceedings of the IEEE Conference on Computer Vision and Pattern Recognition (CVPR 2010), pp. 3424–3431 (2010)
15. Bawa, M., Condie, T., Ganesan, P.: LSH forest: self-tuning indexes for similarity search. In: Proceedings of the International World Wide Web Conference (WWW 2005), pp. 651–660 (2005)
16. Gong, Y., Lazebnik, S.: Iterative quantization: a procrustean approach to learning binary codes. In: CVPR (2011)
17. He, J., Radhakrishnan, R., Chang, S.-F., Bauer, C.: Compact hashing with joint optimization of search accuracy and time. In: CVPR (2011)
18. Joly, A., Buissonr, O.: Random maximum margin hashing. In: CVPR (2011)
19. Weiss, Y., Torralba, A., Fergus, R.: Spectral hashing. In: NIPS (2008)

20. Heo, J.P., Lee, Y., He, J., Chang, S.F., Yoon, S.E.: Spherical hashing: binary code embedding with hyperspheres. IEEE Trans. Pattern Anal. Mach. Intell. **37**, 2304–2316 (2015)
21. Alhwarin, F., Ferrein, A., Scholl, I.: CRVM: circular random variable-based matcher, a novel hashing method for fast NN search in high-dimensional spaces. In: Proceedings of the 7th International Conference on Pattern Recognition Applications and Methods (ICPRAM 2018), pp. 214–221 (2018)
22. Fisher, N.I., Lee, A.: A correlation coefficient for circular data. Biometrika **70**(2), 327–332 (1983)
23. Philbin, J., Chum, O., Isard, M., Sivic, J., Zisserman, A.: Object retrieval with large vocabularies and fast spatial matching. In: Proceedings of the IEEE Conference on Computer Vision and Pattern Recognition (CVPR 2007) (2007)
24. Torralba, A., Fergus, R., Freeman, W.T.: 80 million tiny images: a large data set for nonparametric object and scene recognition. IEEE Trans. Pattern Anal. Mach. Intell. **30**(11), 1958–1970 (2008)

Stochastic Analysis of Time-Difference and Doppler Estimates for Audio Signals

Gabrielle Flood$^{(\boxtimes)}$, Anders Heyden, and Kalle Åström

Centre for Mathematical Sciences, Lund University, Lund, Sweden
{gabrielle.flood,anders.heyden,kalle.astrom}@math.lth.se

Abstract. Pairwise comparison of sound and radio signals can be used to estimate the distance between two units that send and receive signals. In a similar way it is possible to estimate differences of distances by correlating two received signals. There are essentially two groups of such methods, namely methods that are robust to noise and reverberation, but give limited precision and sub-sample refinements that are more sensitive to noise, but also give higher precision when they are initialized close to the real translation. In this paper, we present stochastic models that can explain the precision limits of such sub-sample time-difference estimates. Using these models new methods are provided for precise estimates of time-differences as well as Doppler effects. The developed methods are evaluated and verified on both synthetic and real data.

Keywords: Time-difference of arrival · Sub-sample methods
Doppler effect · Uncertainty measure

1 Introduction

Audio and radio sensors are increasingly used in smartphones, tablet PC's, laptops and other everyday tools. They also form the core of internet-of-things, e.g. small low-power units that can run for years on batteries or use energy harvesting to run for extended periods of time. If the locations of the sensing units are known, they can be used as an ad-hoc acoustic or radio sensor network. There are several interesting cases where such sensor networks can come into use. One such application is localization, cf. [5–7,9]. Another possible usage is beam-forming, i.e. to improve sound quality, [2]. Using a sensor network one can also determine who spoke when through speaker diarisation, [1]. If the sensor positions are unknown or if they are only known to a certain accuracy, the performance of such use-cases are inferior as is shown in [18]. It is, however, possible to perform automatic calibration, i.e. to estimate both sender and receiver positions, even without any prior information, as illustrated in Figs. 1 and 2. This can be done up to a choice of coordinate system, [8,12,13,19,22], thus providing accurate sensor positions for improved use. A key component for all of these methods is the process of obtaining and assessing estimates of e.g. time-difference of arrival

© Springer Nature Switzerland AG 2019
M. De Marsico et al. (Eds.): ICPRAM 2018, LNCS 11351, pp. 116–138, 2019.
https://doi.org/10.1007/978-3-030-05499-1_7

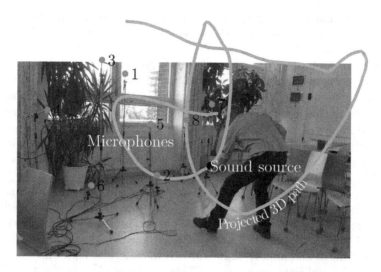

Fig. 1. Precise time-difference of arrival estimation can be used for many purposes, e.g. diarization, beam-forming, positioning and anchor free node calibration. The figure illustrates its use for anchor free node calibration, sound source movement and room reconstruction. The image is taken from [10].

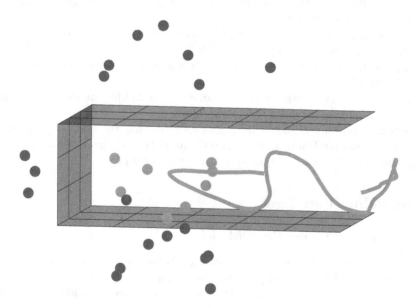

Fig. 2. The figure examplifies one usage of precise time-difference of arrival estimation. The image illustrates the estimated microphone positions (dots), estimated mirrored microphone positions (dots) and sound source motion (solid curve) from Fig. 1. The estimated reflective planes are also shown in the figure. These three planes correspond to the floor, the ceiling and the wall. The image is taken from [10].

of transmitted signals as they arrive in pairs of sensors. In this paper the focus is primarily on acoustic signals, but the same principles are useful for the analysis of radio signals [4].

All of these applications depend on accurate methods to extract features from the sound (or radio) signals. The most common feature is the time-difference-of-arrival, which is then used for subsequent processing. For applications, it is important to find as precise estimates as possible. In [23] time-difference estimates were improved using sub-sample methods. It was also shown empirically that the estimates of the receiver-sender configurations were improved by this. However, no analysis of the uncertainties of the sub-sample time-differences was provided.

This paper is an extended version of [10]. The main content is thus similar. However this version has been developed and is more thorough. E.g. the derivations in Sect. 3.1 have been extended, a comparison between different models has been added, see Sects. 3 and 4.1, and the experiments on real data in Sect. 4.2 have been changed and improved. In addition we have also performed stochastic analysis for the real data experiments. This is presented in Sect. 4.2. Then follows Sect. 4.3 which is partly new. Furthermore, most of the figures have been updated, even if a few remain from the original paper.

The main contributions of [10] and this paper are:

- A scheme for computing time-difference estimates and for estimating the precision of these estimates.
- A method to estimate minute Doppler effects, which is motivated by an experimental comparison between the models.
- An extension of the framework to capture and estimate amplitude differences in the signals.
- An evaluation on synthetic data to evince the validity of the models and provide knowledge of when the method fails.
- An evaluation on real data which demonstrates that the estimates for time-difference, minute Doppler effects and the amplitude changes contain relevant information. This is shown for speeds as small as 0.1 m/s.

2   Modeling Paradigm

2.1   Measurement and Error Model

In this paper, discretely sampled signals are studied. These could e.g. be audio or radio signals. Here, the sampling rate is assumed to be known and constant. Furthermore, we assume that the measured signal y has been ideally sampled after which noise – e.g. from the receivers – has been added, s.t.

$$y(k) = Y(k) + e(k).\tag{1}$$

The original, continuous signal is denoted $Y : \mathbb{R} \mapsto \mathbb{R}$ and the noise, which is a discrete stationary stochastic process, is denoted e.

Let the set of functions $Y : \mathbb{R} \to \mathbb{R}$ that are (i) continuous (ii) square integrable and (iii) with a Fourier transform equal to zero outside $[-\pi, \pi]$ be denoted $\mathbb{B}$. Furthermore, denote the set of discrete, square integrable functions $y : \mathbb{Z} \to \mathbb{R}$ by ℓ. Now, define the discretization operator $D : \mathbb{B} \to \ell$ by

$$y(i) = D(Y)(i) = Y(i). \tag{2}$$

Moreover, we introduce the interpolation operator $I_g : \ell \to \mathbb{B}$, as

$$Y(x) = I_g(y)(x) = \sum_{i=-\infty}^{\infty} g(x - i)y(i). \tag{3}$$

It has been shown that interpolation using the normalized sinc function, i.e. with $g(x) = \mathrm{sinc}(x)$, restores a sampled function for functions in $\mathbb{B}$, see [20] Thus, we call $I_{\mathrm{sinc}} : \ell \to \mathbb{B}$ the ideal interpolation operator and we have that

$$I_{\mathrm{sinc}}(D(Y)) = Y. \tag{4}$$

In the same way other interpolation methods can be expressed similarly. E.g. we obtain Gaussian interpolation by changing sinc in the expression above to

$$G_a(x) = \frac{1}{\sqrt{2\pi a^2}} e^{x^2/(2a^2)}. \tag{5}$$

2.2 Scale-Space Smoothing and Ideal Interpolation

A measured and interpolated signal is often smoothed for two reasons. Firstly, there is often more signal as compared to noise for lower frequencies, whereas for higher frequencies there is usually less signal in relation to noise. Therefore smoothing can be used in order to remove some of the noise, while keeping most of the signal.

Secondly, patterns in a more coarse scale are easier captured after smoothing has been applied, [15]. A Gaussian kernel G_{a_2}, with standard deviation a_2, has been used for the smoothing. We will also refer to a_2 as the *smoothing parameter*.

Given a sampled signal y, the ideally interpolated and smoothed signal can be written as

$$Y(x) = (G_{a_2} * I_{\mathrm{sinc}}(y))(x) = I_{G_{a_2} * \mathrm{sinc}}(y)(x). \tag{6}$$

If a_2 is large enough the approximation $G_{a_2} * \mathrm{sinc} \approx G_{a_2}$ holds. Thus, one can use interpolation with the Gaussian kernel as an approximation for ideal interpolation followed by Gaussian smoothing, [3], s.t.

$$Y(x) = I_{G_{a_2} * \mathrm{sinc}}(y)(x) \approx I_{G_{a_2}}(y)(x). \tag{7}$$

What *large enough* means will be studied in Sect. 4.1.

Moreover, we will later use the fact that discrete w.s.s. Gaussian noise interpolates to continuous w.s.s. Gaussian noise, as is shown in [3].

3   Time-Difference and Doppler Estimation

Assume that we have two signals, $W(t)$ and $\bar{W}(t)$. The signals are measured and interpolated as described above. Also assume that the two signals are similar, but with one e.g. translated and compressed in the time domain. This could occur when two different receivers pick up an audio signal from a single sender. Then the second signal can be obtained from the other and a few parameters. We describe the relation as follows

$$W(t) = \bar{W}(\alpha t + h), \tag{8}$$

where h describes the time-difference of arrival, or translation in the signals. In a setup where the sound source has equal distance to both microphones $h = 0$. The second parameter, α, is a Doppler factor. This parameter is needed for example if the sound source or the microphones are moving. For a stationary setup $\alpha = 1$.

When the two microphones pick up the signals these are disturbed by Gaussian w.s.s. noise. Thus, the received signals can be written

$$V(t) = W(t) + E(t) \quad \text{and} \quad \bar{V}(t) = \bar{W}(t) + \bar{E}(t). \tag{9}$$

Here, $E(t)$ and $\bar{E}(t)$ denotes the two independent noise signals after interpolation.

Assume that the signals V and $\bar{V}$ are given. Also, denote by $z = \begin{bmatrix} z_1 & z_2 \end{bmatrix}^T = \begin{bmatrix} h & \alpha \end{bmatrix}^T$, the vector of unknown parameters. Then, the parameters for which (8) is true can be estimated by the z that minimizes the integral

$$G(z) = \int_t (V(t) - \bar{V}(z_2 t + z_1))^2 \, dt. \tag{10}$$

Comparing with Cross Correlation. If we only estimate a time delay h, the minimization of the error function (10) would in practice be the same as maximizing the cross correlation of V and $\bar{V}$. The cross-correlation for real signals is defined as

$$(V \star \bar{V})(h) = \int_t V(t)\bar{V}(t + h) \, dt. \tag{11}$$

Thus, the h that maximize this cross-correlation is given by

$$\text{argmax}_h (V \star \bar{V})(h) = \text{argmax}_h \int_t V(t)\bar{V}(t + h) \, dt. \tag{12}$$

If we expand the error function (10), while neglecting the Doppler factor we obtain the minimizer

$$\text{argmin}_h \int_t (V(t) - \bar{V}(t + h))^2 \, dt = \text{argmin}_h \int_t (V(t))^2 + (V(t+h))^2 - 2V(t)\bar{V}(t + h) \, dt$$

$$= \text{argmin}_h \int_t -2V(t)\bar{V}(t + h) \, dt = \text{argmax}_h \int_t V(t)\bar{V}(t + h) \, dt. \tag{13}$$

Note that since we integrate over t, the integral $\int_t (\bar{V}(t+h))^2\, dt$ is almost constant, ignoring edge effects.

We choose to use (10) for estimation of the parameters since it is simple to expand and is valid even if we add more parameters.

3.1 Estimating the Standard Deviation of the Parameters

If $\boldsymbol{z}_T = \begin{bmatrix} h_T & \alpha_T \end{bmatrix}^T$ is the "true" parameter for the data and $\hat{\boldsymbol{z}}$ is the parameter that has been estimated by minimizing (10), the estimation error can be expressed as

$$X = \hat{\boldsymbol{z}} - \boldsymbol{z}_T. \tag{14}$$

Assume, without loss of generality, that $\boldsymbol{z}_T = \begin{bmatrix} 0 & 1 \end{bmatrix}^T$. The standard deviation of $\hat{\boldsymbol{z}}$ will be the same as the standard deviation of X and the mean of those two will only differ by $\boldsymbol{z}_T$. Thus, it is sufficient to study X to get statistical information about the estimate $\hat{\boldsymbol{z}}$.

Linearizing $G(\boldsymbol{z})$ around the true displacement $\boldsymbol{z}_T = \begin{bmatrix} 0 & 1 \end{bmatrix}^T$ gives

$$G(\boldsymbol{z}) \approx F(X) = \frac{1}{2} X^T a X + b X + f, \tag{15}$$

with

$$a = \nabla^2 G(\boldsymbol{z}_T), \qquad b = \nabla G(\boldsymbol{z}_T), \qquad f = G(\boldsymbol{z}_T). \tag{16}$$

Using (9) and (8), we get

$$f = G([01]^T) = \int_t (V(t) - \bar{V}(1 \cdot t - 0))^2\, dt = \int_t (W(t) + E(t) - (\bar{W}(t) + \bar{E}(t)))^2\, dt$$
$$= \int_t (E - \bar{E})^2\, dt = \int_t E^2 + 2E\bar{E} + \bar{E}^2\, dt. \tag{17}$$

To find the coefficients a and b we first calculate the derivatives $\nabla G(\boldsymbol{z})$ and $\nabla^2 G(\boldsymbol{z})$.

$$\nabla G = \begin{bmatrix} \int_t 2(V(t) - \bar{V}(\alpha t + h)) \cdot (-\bar{V}'(\alpha t + h))\, dt \\ \int_t 2(V(t) - \bar{V}(\alpha t + h)) \cdot (-\bar{V}'(\alpha t + h) \cdot t)\, dt \end{bmatrix}$$
$$= -2 \begin{bmatrix} \int_t (W(t) + E(t) - \bar{W}(\alpha t + h) - \bar{E}(\alpha t + h)) \cdot (\bar{W}'(\alpha t + h) + \bar{E}'(\alpha t + h))\, dt \\ \int_t (W(t) + E(t) - \bar{W}(\alpha t + h) - \bar{E}(\alpha t + h)) \cdot (\bar{W}'(\alpha t + h) + \bar{E}'(\alpha t + h)) \cdot t\, dt \end{bmatrix}. \tag{18}$$

Inserting the true displacement $\boldsymbol{z}_T$, at the point of linearization, gives

$$b = \nabla G(\boldsymbol{z}_T)$$
$$= -2 \begin{bmatrix} \int_t (W(t) + E(t) - \bar{W}(t) - \bar{E}(t)) \cdot (\bar{W}'(t) + \bar{E}'(t))\, dt \\ \int_t (W(t) + E(t) - \bar{W}(t) - \bar{E}(t)) \cdot (\bar{W}'(t) + \bar{E}'(t)) \cdot t\, dt \end{bmatrix}$$
$$= -2 \begin{bmatrix} \int_t (E - \bar{E})(\bar{W}' + \bar{E}')\, dt \\ \int_t (E - \bar{E})(\bar{W}' + \bar{E}')t\, dt \end{bmatrix} = -2 \begin{bmatrix} \int_t E\bar{W}' + E\bar{E}' - \bar{E}\bar{W}' - \bar{E}\bar{E}'\, dt \\ \int_t (E\bar{W}' + E\bar{E}' - \bar{E}\bar{W}' - \bar{E}\bar{E}')t\, dt \end{bmatrix}. \tag{19}$$

To simplify further computations, we introduce

$$\hat{\varphi} = E\bar{W}' + E\bar{E}' - \bar{E}\bar{W}' - \bar{E}\bar{E}', \tag{20}$$

such that

$$b = -2\begin{bmatrix} \int_t \hat{\varphi}\, dt \\ \int_t t\hat{\varphi}\, dt \end{bmatrix}. \tag{21}$$

Furthermore,

$$\nabla^2 G =$$
$$\begin{bmatrix} \int_t -2\bar{V}'(\alpha t + h) \cdot (-\bar{V}'(\alpha t + h)) + 2(V(t) - \bar{V}(\alpha t + h))(-\bar{V}''(\alpha t + h))\, dt \\ \int_t -2\bar{V}'(\alpha t + h) \cdot t \cdot (-\bar{V}'(\alpha t + h)) + 2(V(t) - \bar{V}(\alpha t + h))(-\bar{V}''(\alpha t + h) \cdot t)\, dt \end{bmatrix} \cdots$$
$$\begin{bmatrix} \int_t -2\bar{V}'(\alpha t + h) \cdot (-\bar{V}'(\alpha t + h)) \cdot t + 2(V(t) - \bar{V}(\alpha t + h)) \cdot (-\bar{V}''(\alpha t + h) \cdot t)\, dt \\ \int_t -2\bar{V}'(\alpha t + h) \cdot t(-\bar{V}'(\alpha t + h)) \cdot t + 2(V(t) - \bar{V}(\alpha t + h)) \cdot (-\bar{V}''(\alpha t + h) \cdot t^2)\, dt \end{bmatrix}$$
$$= 2\begin{bmatrix} \int_t (\bar{V}'(\alpha t + h))^2 - V(t)\bar{V}''(\alpha t + h) + \bar{V}(\alpha t + h)\bar{V}''(\alpha t + h)\, dt \\ \int_t t \cdot (\bar{V}'(\alpha t + h))^2 - t \cdot V(t)\bar{V}''(\alpha t + h) + t \cdot \bar{V}(\alpha t + h)\bar{V}''(\alpha t + h)\, dt \end{bmatrix} \cdots$$
$$\begin{bmatrix} \int_t \cdot(\bar{V}'(\alpha t + h))^2 - t \cdot V(t)\bar{V}''(\alpha t + h) + t \cdot \bar{V}(\alpha t + h)\bar{V}''(\alpha t + h)\, dt \\ \int_t t^2 \cdot (\bar{V}'(\alpha t + h))^2 - t^2 \cdot V(t)\bar{V}''(\alpha t + h) + t^2 \cdot \bar{V}(\alpha t + h)\bar{V}''(\alpha t + h) \end{bmatrix}. \tag{22}$$

Now, introducing the notation

$$\phi(z) = (\bar{V}'(\alpha t + h))^2 - V(t)\bar{V}''(\alpha t + h) + \bar{V}(\alpha t + h)\bar{V}''(\alpha t + h), \tag{23}$$

we can write $\nabla^2 G$ shorter as

$$\nabla^2 G = \begin{bmatrix} \int_t \phi\, dt & \int_t t\phi\, dt \\ \int_t t\phi\, dt & \int_t t^2\phi\, dt \end{bmatrix}. \tag{24}$$

If we let $\hat{\phi}$ be the value of ϕ for z_T

$$\hat{\phi} = \phi(z_T) = (\bar{W}'(t) + \bar{E}'(t))^2 - (W(t) + E(t))(\bar{W}''(t) + \bar{E}''(t))$$
$$+ (\bar{W}(t) + \bar{E}(t))(\bar{W}''(t) + \bar{E}''(t)) \tag{25}$$
$$= (\bar{W}')^2 + 2\bar{W}'\bar{E}' + (\bar{E}')^2 - E\bar{W}'' - E\bar{E}'' + \bar{E}\bar{W}'' + \bar{E}\bar{E}''$$

we get

$$a = \nabla^2 G(z_T) = \begin{bmatrix} \int_t \hat{\phi}\, dt & \int_t t\hat{\phi}\, dt \\ \int_t t\hat{\phi}\, dt & \int_t t^2\hat{\phi}\, dt \end{bmatrix}. \tag{26}$$

We also have that $F(X) = 1/2 \cdot X^T a X + bX + f$. To minimize this error function, we find the X for which the derivative of $F(X)$ is zero. Since a is symmetric we get

$$\nabla F(X) = aX + b = 0 \quad \Leftrightarrow \quad X = g(a, b) = -a^{-1}b. \tag{27}$$

In the calculations below, we assume that a is invertible.

Now we would like to find the mean and covariance of X. For this, Gauss' approximation formulas are used. If we denote the expected value of a and b with $\mu_a = \mathbf{E}[A]$ and $\mu_b = \mathbf{E}[b]$ respectively the expected value of X can be approximated to

$$
\begin{aligned}
\mathbf{E}[X] &= \mathbf{E}[g(a,b)] \approx \mathbf{E}[g(\mu_a, \mu_b) + (a - \mu_a)g_a'(\mu_a, \mu_b) + (b - \mu_b)g_b'(\mu_a, \mu_b)] \\
&= g(\mu_a, \mu_b) + (\mathbf{E}[a] - \mu_a)g_a'(\mu_a, \mu_b) + (\mathbf{E}[b] - \mu_b)g_b'(\mu_a, \mu_b) \\
&= g(\mu_a, \mu_b) = -\mu_a^{-1}\mu_b = -\mathbf{E}[a]^{-1}\mathbf{E}[b].
\end{aligned}
\tag{28}
$$

In a similar manner the covariance of X is

$$
\begin{aligned}
\mathbf{C}[X] &= \mathbf{C}[g(a,b)] \approx g_a'(\mu_a, \mu_b)\mathbf{C}[a]g_a'(\mu_a, \mu_b)^T + g_b'(\mu_a, \mu_b)\mathbf{C}[b]g_b'(\mu_a, \mu_b)^T \\
&\quad + 2g_a'(\mu_a, \mu_b)\mathbf{C}[a, b]g_b'(\mu_a, \mu_b)^T,
\end{aligned}
\tag{29}
$$

where $\mathbf{C}[a, b]$ denotes the cross-covariance between a and b. For further computations $g_a'(a, b)$, $g_b'(a, b)$, $\mathbf{E}[a]$, $\mathbf{E}[b]$, $\mathbf{C}[b]$ and $\mathbf{C}[a, b]$ are needed.

By computing the expected value of $\hat{\varphi}$

$$
\begin{aligned}
\mathbf{E}[\hat{\varphi}] &= \mathbf{E}[E\bar{W}' + E\bar{E}' - \bar{E}\bar{W}' - \bar{E}\bar{E}'] \\
&= \mathbf{E}[E]\bar{W}' + \mathbf{E}[E]\mathbf{E}[\bar{E}'] - \mathbf{E}[\bar{E}]\bar{W}' - \mathbf{E}[\bar{E}]\mathbf{E}[\bar{E}'] = 0
\end{aligned}
\tag{30}
$$

we get

$$
\mathbf{E}[b] = \mathbf{E}\left[-2 \begin{bmatrix} \int_t \hat{\varphi}\,dt \\ \int_t t\hat{\varphi}\,dt \end{bmatrix} \right] = -2 \begin{bmatrix} \int_t \mathbf{E}[\hat{\varphi}]\,dt \\ \int_t t\mathbf{E}[\hat{\varphi}]\,dt \end{bmatrix} = -2 \begin{bmatrix} \int_t 0\,dt \\ \int_t t \cdot 0\,dt \end{bmatrix} = \begin{bmatrix} 0 \\ 0 \end{bmatrix}.
\tag{31}
$$

In the second step of the computation of $\mathbf{E}[\hat{\varphi}]$ we have used the fact that for a weakly stationary process the process and its derivative at a certain time are uncorrelated, and thus $\mathbf{E}[\bar{E}\bar{E}'] = \mathbf{E}[\bar{E}]\mathbf{E}[\bar{E}']$, [16]. Hence,

$$
\mathbf{E}[X] = -\mathbf{E}[a]^{-1}\mathbf{E}[b] = -\mathbf{E}[a]^{-1}\begin{bmatrix} 0 \\ 0 \end{bmatrix} = \begin{bmatrix} 0 \\ 0 \end{bmatrix}.
\tag{32}
$$

For the partial derivative of $g(a, b)$ w.r.t. b we get [17]

$$
g_b'(a, b) = \frac{\partial}{\partial b}\left(-a^{-1}b \right) = -(a^{-1})^T = -(a^T)^{-1} = -a^{-1}
\tag{33}
$$

and thus $g_b'(\mu_a, \mu_b) = -(\mathbf{E}[a])^{-1}$. Since $\mathbf{E}[b] = 0$, we get that $g_a'(\mu_a, \mu_b) = 0$, [17]. Hence the first and the last term in (29) cancel, leaving

$$
\begin{aligned}
\mathbf{C}[X] &= g_b'(\mu_a, \mu_b)\mathbf{C}[b]g_b'(\mu_a, \mu_b)^T = (-\mathbf{E}[a]^{-1})\mathbf{C}[b](-\mathbf{E}[a]^{-1})^T \\
&= \mathbf{E}[a]^{-1}\mathbf{C}[b](\mathbf{E}[a]^{-1})^T.
\end{aligned}
\tag{34}
$$

To find the expected value of a the expected value of $\hat{\phi}$ is needed. This is obtained from

$$\mathbf{E}[\hat{\phi}] = (\bar{W}')^2 + 2\bar{W}'\mathbf{E}[\bar{E}'] + \mathbf{E}[(\bar{E}')^2] - \bar{W}''\mathbf{E}[E] - \mathbf{E}[E]\mathbf{E}[\bar{E}''] + \bar{W}''\mathbf{E}[\bar{E}] + \mathbf{E}[\bar{E}\bar{E}'']$$
$$= (\bar{W}')^2 + \mathbf{E}[(\bar{E}')^2] + \mathbf{E}[\bar{E}\bar{E}''] = (\bar{W}')^2. \tag{35}$$

In the last equality we have used that $\mathbf{E}[\bar{E}\bar{E}''] = -\mathbf{E}[(\bar{E}')^2]$, [16]. Thus, the two last terms cancel out. The expected value of a is therefore

$$\mathbf{E}[a] = 2\begin{bmatrix} \int_t \mathbf{E}[\hat{\phi}]\,dt & \int_t \mathbf{E}[t\hat{\phi}]\,dt \\ \int_t \mathbf{E}[t\hat{\phi}]\,dt & \int_t \mathbf{E}[t^2\hat{\phi}]\,dt \end{bmatrix} = 2\begin{bmatrix} \int_t (\bar{W}')^2\,dt & \int_t t(\bar{W}')^2\,dt \\ \int_t t(\bar{W}')^2\,dt & \int_t t^2(\bar{W}')^2\,dt \end{bmatrix}. \tag{36}$$

Now, since the expected value of b is zero, the covariance of b is

$$\mathbf{C}[b] = (-2)^2\begin{bmatrix} C_{11} & C_{12} \\ C_{21} & C_{22} \end{bmatrix}, \tag{37}$$

with

$$C_{11} = \mathbf{E}\left[\int_{t_1} \hat{\varphi}(t_1)\,dt_1 \cdot \int_{t_2} \hat{\varphi}(t_2)\,dt_2\right]$$

$$C_{12} = \mathbf{E}\left[\int_{t_1} t_1\hat{\varphi}(t_1)\,dt_1 \cdot \int_{t_2} \hat{\varphi}(t_2)\,dt_2\right]$$

$$C_{21} = \mathbf{E}\left[\int_{t_1} \hat{\varphi}(t_1)\,dt_1 \cdot \int_{t_2} t_2\hat{\varphi}(t_2)\,dt_2\right] \tag{38}$$

$$C_{22} = \mathbf{E}\left[\int_{t_1} t_1\hat{\varphi}(t_1)\,dt_1 \cdot \int_{t_2} t_2\hat{\varphi}(t_2)\,dt_2\right].$$

Note that by changing the order of the terms in C_{12} it is clear that $C_{21} = C_{12}$. Furthermore, we obtain

$$C_{11} = \mathbf{E}\left[\int_{t_1} \hat{\varphi}(t_1)\,dt_1 \cdot \int_{t_2} \hat{\varphi}(t_2)\,dt_2\right]$$

$$= \mathbf{E}\left[\left(\int_{t_1} (E - \bar{E})(\bar{W}' + \bar{E}')\,dt_1\right) \cdot \left(\int_{t_2} (E - \bar{E})(\bar{W}' + \bar{E}')\,dt_2\right)\right] \tag{39}$$

$$= \mathbf{E}\left[\int_{t_1}\int_{t_2} (E(t_1) - \bar{E}(t_1))(\bar{W}'(t_1) + \bar{E}'(t_1))\cdot\right.$$
$$\left.(E(t_2) - \bar{E}(t_2))(\bar{W}'(t_2) + \bar{E}'(t_2))\,dt_1dt_2\right].$$

Denoting $\mathbf{E}[(E(t_1) - \bar{E}(t_1))(E(t_2) - \bar{E}(t_2))] = r_{E-\bar{E}}(t_1 - t_2)$ and assuming that $\mathbf{E}[\bar{E}'(t_1)\bar{E}'(t_2)]$ is small gives

$$C_{11} = \mathbf{E}\left[\int_{t_1} \hat{\varphi}(t_1)\, dt_1 \cdot \int_{t_2} \hat{\varphi}(t_2)\, dt_2\right]$$

$$= \int_{t_1}\int_{t_2} \mathbf{E}[(E(t_1) - \bar{E}(t_1))(E(t_2) - \bar{E}(t_2))] \cdot (\bar{W}'(t_1)\bar{W}'(t_2)$$

$$\qquad + \bar{W}'(t_1)\mathbf{E}[\bar{E}'(t_2)] + \mathbf{E}[\bar{E}'(t_1)]\bar{W}'(t_2) + \mathbf{E}[\bar{E}'(t_1)\bar{E}'(t_2)])\, dt_2 dt_1 \qquad (40)$$

$$\approx \int_{t_1}\int_{t_2} r_{E-\bar{E}}(t_1 - t_2)\bar{W}'(t_1)\bar{W}'(t_2)\, dt_2 dt_1$$

$$= \int_{t_1} \bar{W}'(t_1)(\bar{W}' * r_{E-\bar{E}})(t_1)\, dt_1.$$

The time t is a deterministic quantity and the other elements in $\mathbf{C}[b]$ can be computed similarly. Finally we have

$$C_{11} = \int_t \bar{W}'(t)(\bar{W}' * r_{E-\bar{E}})(t)\, dt$$

$$C_{12} = C_{21} = \int_t t\bar{W}'(t)(\bar{W}' * r_{E-\bar{E}})(t)\, dt \qquad (41)$$

$$C_{22} = \int_t t\bar{W}'(t)((t\bar{W}') * r_{E-\bar{E}})(t)\, dt$$

and through (34) we get an expression for the variance and thus also the standard deviation of X.

3.2 Expanding the Model

It is easy to change or expand the model (8) to contain more (or fewer) parameters. If we keep h and α and add an extra amplitude parameter γ, we get the model

$$W(t) = \gamma \bar{W}(\alpha t + h). \qquad (42)$$

The error integral (10) would then be changed accordingly and the optimization would instead be over over $z = \begin{bmatrix} z_1 & z_2 & z_3 \end{bmatrix} = \begin{bmatrix} h & \alpha & \gamma \end{bmatrix}$.

The computations for achieving the estimations does in practice not get harder when we add more parameters. However, the analysis from the previous section gets more complex.

4 Experimental Validation

For validation we perform experiments on both real data and synthetic data. The purpose of using synthetic data is to demonstrate the validity of the model, but also to verify the approximations used. In the latter case we have studied at what signal-to-noise ratio the approximations are valid. Furthermore, to show that the parameter estimations contain useful information, we have done experiments on real data. This is well-known for time-difference, but less explored for the Doppler effects and amplitude changes.

4.1   Synthetic Data - Validation of Method

The model was first tested on simulated data in order to study when the approximations in the model derivation hold. The linearization using Gauss' approximation formula, e.g. (28) and (29), is one example of such approximations. Another is the usage of Gaussian interpolation as an approximation of ideal interpolation followed by convolution with a Gaussian, (7).

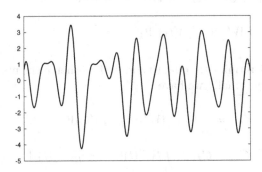

Fig. 3. The simulated signal that was used for the experimental validation. To achieve a more realistic signal noise of different levels was added later on. The plot is taken from [10].

To do these studies we compared the theoretical standard deviations of the parameters calculated according to Sect. 3.1 with empirically computed standard deviations. The agreement of these standard deviations makes us conclude that our approximations are valid.

First we simulated an original continuous signal $W(x)$, see Fig. 3. The second signal was then created according to (8) s.t. $\bar{W} = W(1/\alpha \cdot (x - h))$. The signals were ideally sampled after which Gaussian white discrete noise with standard deviation σ_n was added. After smoothing with a Gaussian kernel with standard deviation a_2 (see Sect. 2.2) the signals can be described by $V(t)$ and $\bar{V}(t)$ as before.

The two signals V and $\bar{V}$ were simulated anew 1000 times to investigate the effect of a_2 and σ_n. Each time the same original signals W and $\bar{W}$ were used, but with different noise realizations. Then, we computed the theoretical standard deviation of the parameter vector z, $\sigma_z = \begin{bmatrix} \sigma_h & \sigma_\alpha \end{bmatrix}$. This was done in accordance with the presented theory. We also computed an empirical standard deviation $\hat{\sigma}_z = \begin{bmatrix} \hat{\sigma}_h & \hat{\sigma}_\alpha \end{bmatrix}$ from the 1000 different parameter estimations.

When studying the effect of a_2 the noise level was kept constant, with $\sigma_n = 0.03$. The translation was set to $h = 3.63$ and the Doppler factor was $\alpha = 1.02$. However, the exact numbers are unessential. While varying the smoothing parameter $a_2 \in [0.3, 0.8]$ the standard deviation was then computed according to the procedure above.

The results from these simulations can be seen in Fig. 4. When a_2 is below $a_2 \approx 0.55$ the theoretical values σ_z and the empirical values $\hat{\sigma}_z$ do not agree,

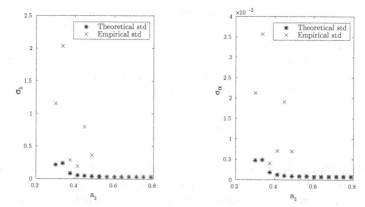

Fig. 4. The plots show the standard deviation of the parameters in z for different values of the smoothing parameter a_2. The stars ($*$) represent the theoretical values σ_z and the crosses (x) the empirical values $\hat{\sigma}_z$. The left plot shows the results for the translation $z_1 = h$ and the right plot for the Doppler factor $z_2 = \alpha$. It is clear that the approximation is valid approximately when $a_2 > 0.55$. The plots are taken from [10].

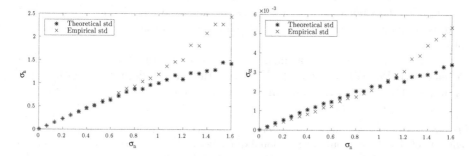

Fig. 5. The standard deviation of the translation (to the left) and Doppler factor (to the right) for different levels of noise in the signal. The stars ($*$) mark the theoretical values σ_z and the crosses (x) the empirical $\hat{\sigma}_z$. For the translation the values agree for signals with a noise level up to $\sigma_n \approx 0.8$. For the Doppler factor the theoretical values follow the empirical values when $\sigma_n < 1.1$. The plots are taken from [10].

while they do for $a_2 > 0.55$. Therefore we draw the conclusion that the approximation (7) of ideal interpolation should only be used when $a_2 > 0.55$.

Secondly, the effect of changing the noise level was investigated. The smoothing parameters was fixed to $a_2 = 2$ and the translation and the Doppler factor were kept on the same level as before. Instead we varied the noise level s.t. $\sigma_n \in [0, 1.6]$. Then the standard deviations of the parameters σ_z and $\hat{\sigma}_z$ were computed in the same way as in the previous section.

The results from this run can be seen in Fig. 5, with the results for the translation parameter h to the left and for the Doppler parameter α to the right. When σ_n is lower than $\sigma_n \approx 0.8$ the theoretical and empirical values for the translation parameter are similar. For higher values of σ_n they do not agree. The same goes for the Doppler factor when the noise level is below $\sigma_n \approx 1.1$.

By this, we reason that noise with a standard deviation up to $\sigma_n \approx 0.8$ can be handled. The original signal W have an amplitude that varies between 1 and 3.5 and using the standard deviation of that signal, σ_W, we can compute the signal-to-noise ratio that the system can manage. We get the result

$$\text{SNR} = \frac{\sigma_W^2}{\sigma_n^2} \approx 4.7. \tag{43}$$

Comparing Different Models. In this paper we have chosen to work with the models (8) and (42). However, we have so far not presented any comparison between different models. To investigate this, we studied two models, namely (8), which we call model B and a slightly simpler model which we call model A,

$$W(x) = \bar{W}(x + h). \tag{44}$$

To begin with, we simulated data according to model A. We call this data A. During the simulation the standard deviation of the noise in the signals was set to $\sigma_n = 0.02$ and the smoothing parameter was $a_2 = 2.0$. Furthermore, we studied this data both using model A, i.e. by minimizing $\int_t (V(t) - \bar{V}(t+h))^2 \, dt$ and using model B, see (10). The results can be seen in the first column (Data A) of Table 1.

Secondly, a similar test was made but this time we simulated data according to model B. We call this data B. We then studied this data using both model A and B. The results are shown in the second column (data B) of Table 1.

Table 1. Comparison between model A from (44) and model B from (8). Data A consists of signals with only translational differences while the second signal in data B is affected by both translation and a Doppler effect. The standard deviations for model B in the table regards the theoretical values that were derived in Section 3.1, and a similar analysis has been performed for model A.

		Data A	Data B
True values	Translation, h_T	3.63	3.63
	Doppler factor, α_T	1.00	1.02
Model A	Est. h, $\hat{h}^{(A)}$	3.63	13.4
	Std. of h, $\sigma_h^{(A)}$	$1.01 \cdot 10^{-2}$	$1.02 \cdot 10^{-2}$
Model B	Est. h, $\hat{h}^{(B)}$	3.63	3.66
	Std. of h, $\sigma_h^{(B)}$	$2.30 \cdot 10^{-2}$	$2.30 \cdot 10^{-2}$
	Est. α, $\hat{\alpha}^{(B)}$	1.00	1.02
	Std. of α, $\sigma_\alpha^{(B)}$	$5.32 \cdot 10^{-5}$	$5.33 \cdot 10^{-5}$

Studying the first column of Table 1 we see that model B estimates the parameters as good as model A – which in this case is the most correct model –

does. Though, for model B the standard deviation σ_h is more than twice as big as for model A.

In the second column of the table we see that since model A cannot estimate the Doppler effect, the translation parameter is erroneously estimated. The standard deviation σ_h is however still lower for model A. To minimize the error function model A estimates the translation such that the signal is fitted in the middle, see Fig. 6. This means that even though the standard deviation is low, the bias is high.

If we know that our collected data has only been affected by a translation it is clearly better to use model A. However, the loss for using a more simple model is larger on complex data than the loss for using a larger model for simple data. Thus, based on the results from Table 1 we conclude that it is better to use a larger model for the real data in the following section.

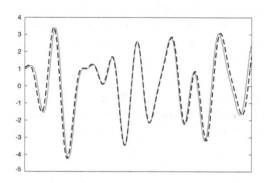

Fig. 6. The results after using model A on data B, where the second signal is affected both by a translation and Doppler effect. Since the model does not estimate any Doppler factor, the estimated translation will be biased. The two signals agree well in the middle, while there is a gap between them at the beginning and the end. This gap cannot be captured by a translation.

4.2 Real Data - Validation of Method

The experiments on real data were performed in an anaechoic chamber and the recording frequency was $f = 96\,\text{kHz}$. We used 8 T-Bone MM-1 microphones and these were connected to an audio interface (M-Audio Fast Track Ultra 8R) and a computer. Furthermore, the microphones were placed so that they spanned 3D, approximately 0.3–1.5 m away from each other. As a sound source we used a mobile phone which was connected to a small loudspeaker. The mobile phone was moved around in the room while playing a song.

We used the technique described in [22] and refined in [21] to achieve ground truth consisting of a 3D trajectory for the sound source path $s(t)$ and the 3D positions of the microphones $r_1, \ldots, r_8$. The method uses RANSAC algorithms which are based on minimal solvers [14] to find initial estimates of the sound trajectory and microphone positions. Then, these are refined using non-linear optimization of a robust error norm, including a smooth motion prior, to reach the final estimates.

However, to make ground truth independent from the data that we used for testing we chose to only take data from microphone 3–8 into account during the first two thirds of the sound signal. Thus, by that we estimated $s(t)$ for certain t and $r_3, \ldots, r_8$. For the final third of the signal we added the information from microphone 1 and 2 as well, such that our solution would not drift compared to ground truth. By that we estimated the rest of $s(t)$, r_1 and r_2.

We only used data from microphone 1 and 2 for the validation of the method presented in this paper. The sound was played for around 29 s and the loudspeaker was constantly moving during this time. Furthermore, both the direction and the speed of the sound source changed.

Since our method assume a constant parameter z in a window we divided the recording into 2834 patches of 1000 samples each (i.e. about 0.01 s). Within these patches the parameters were approximately constant. Each of the patches could then be investigated and compared to ground truth separately. From ground truth we had a constant loudspeaker position $s^{(i)}$, its derivative $\frac{\partial s^{(i)}}{\partial t}(i)$ and the receiver positions r_1 and r_2 for each signal patch i.

Estimating the Parameters. If we call signal patch i from the first microphone $V^{(i)}(t)$ and let $\bar{V}^{(i)}(t)$ be the patch from the second microphone we can estimate the parameters using (8) to model the received signals.

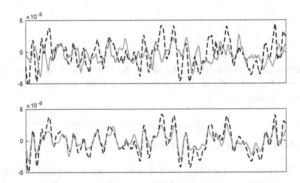

Fig. 7. The received signal patches at a certain time – the first signal in dashed (- -) and the second as solid (   ). The top plot shows the signals as they were received. In the lower plot the same patches have been modified using the optimal parameters h and α.

The method presented in this paper is developed to estimate small translations, s.t. $h \in [-10, 10]$ samples. However, in the experiments the delays were larger than that. Therefore we began by pre-estimating an integer delay $\tilde{h}^{(i)}$ using GCC-PHAT. The GCC-PHAT method is described in [11]. After that we did a subsample refinement of the translation and estimated the Doppler parameter using our method. This was done by minimization of the intergral

$$\int_t (V^{(i)}(t) - \bar{V}^{(i)}(\alpha^{(i)}t + \tilde{h}^{(i)} + h^{(i)}))^2 \, dt. \tag{45}$$

Here, the optimization was over $h^{(i)}$ and $\alpha^{(i)}$, while $\tilde{h}^{(i)}$ should be seen as a constant.

The results after applying the optimized parameters to one of the signal patches can be seen in Fig. 7. The optimization was carried out for all different patches.

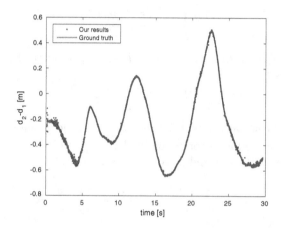

Fig. 8. The figure shows the difference between the distances from receiver 1 to the sender (d_1) and receiver 2 to the sender (d_2) over time. The ground truth $\Delta d^{(i)}$ is plotted as a solid line (   ) and the values $\Delta \bar{d}^{(i)}$ obtained from time-difference estimates as dots (•). Each dot represents the value for one signal patch. It is hard to distinguish the line representing ground truth since the estimations agree well with this. The plot is similar to Fig. 7 in [10], but has been generated using the updated and more independent method which is presented in this paper.

Comparison with Ground Truth. The distances $d_1^{(i)}$ and $d_2^{(i)}$ from the microphones to the loudspeaker were computed from the ground truth receiver and sender positions (r_1, r_2 and $s^{(i)}$) according to

$$d_1^{(i)} = |r_1 - s^{(i)}|, \qquad d_2^{(i)} = |r_2 - s^{(i)}|. \tag{46}$$

The difference of these distances,

$$\Delta d^{(i)} = d_2^{(i)} - d_1^{(i)} \tag{47}$$

has a connection to our estimated translation $h^{(i)}$ and the time difference of arrival. However, $\Delta d^{(i)}$ is measured in meters, while we compute $h^{(i)}$ in samples.

To be able to compare these two, we multiplied $h^{(i)}$ with a scaling factor c/f. The recording frequency was $f = 96$ kHz and $c = 340$ m/s is the speed of sound. From this we could obtain an estimation of $\Delta d^{(i)}$,

$$\Delta \bar{d}^{(i)} = \frac{c}{f} \cdot h^{(i)}. \tag{48}$$

Thereafter we could compare our estimated values $\Delta \bar{d}^{(i)}$ to the ground truth values $\Delta d^{(i)}$. The ground truth is plotted together with our estimations in Fig. 8. The plot shows the results over time, for all different patches. It is clear that the two agree well.

The Doppler parameter measures how the distance differences changes, i.e.

$$\frac{\partial \Delta d}{\partial t} = \frac{\partial d_2}{\partial t} - \frac{\partial d_1}{\partial t}. \tag{49}$$

Here, the distances over time are denoted d_1 and d_2 respectively. The derivative of $d_1(t) = |r_1 - s(t)|$ is

$$\frac{\partial d_1}{\partial t} = \frac{r_1 - s}{|r_1 - s|} \cdot \frac{\partial s}{\partial t}, \tag{50}$$

where $\cdot$ denotes the scalar product between the two time dependent vectors. The derivative of d_2 can be found correspondingly. If $n_1^{(i)}$ and $n_2^{(i)}$ are unit vectors in the direction from $s^{(i)}$ to r_1 and r_2 respectively, i.e.

$$n_1^{(i)} = \frac{r_1 - s^{(i)}}{|r_1 - s^{(i)}|}, \qquad n_2^{(i)} = \frac{r_2 - s^{(i)}}{|r_2 - s^{(i)}|}, \tag{51}$$

the derivatives can be expressed as

$$\frac{\partial d_1^{(i)}}{\partial t} = n_1^{(i)} \cdot \frac{\partial s^{(i)}}{\partial t}, \qquad \frac{\partial d_2^{(i)}}{\partial t} = n_2^{(i)} \cdot \frac{\partial s^{(i)}}{\partial t}. \tag{52}$$

Thus

$$\frac{\partial \Delta d^{(i)}}{\partial t} = n_2^{(i)} \cdot \frac{\partial s^{(i)}}{\partial t} - n_1^{(i)} \cdot \frac{\partial s^{(i)}}{\partial t}. \tag{53}$$

These ground truth Doppler values can be interpreted as how much Δd changes each second. However, our estimated Doppler factor α is a unit-less constant. We can express the relation between the two values as

$$\frac{\partial \Delta d}{\partial t} = (\alpha - 1) \cdot c, \tag{54}$$

where c still denotes the speed of sound. In Fig. 9 the ground truth is plotted as a solid line together with our estimations marked with dots. The similarities are easily distinguishable even if the estimations are noisy.

It is clear from the plots that the estimations contain relevant information. However, there is quite some noise in the estimates in Figs. 8 and 9. This can be reduced further by computation of a moving average. We have computed a

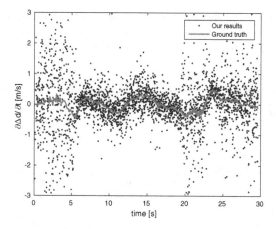

Fig. 9. The derivative of the distance differences Δd plotted over time. The dots (•) are our estimations and the solid line ( ) is computed from ground truth. We see that even though the estimations are noisy the pattern agree with ground truth. The plot is similar to Fig. 8 in [10], but has been generated using the updated and more independent method which is presented in this paper.

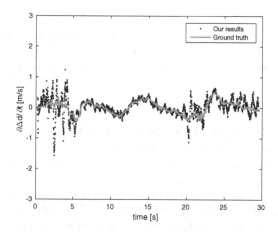

Fig. 10. This plot shows essentially the same thing as Fig. 9, i.e. $\partial \Delta d/\partial t$, but with a 20-patches moving average over the estimations. The averaging substantially reduces the noise. The plot is similar to Fig. 10 in [10], but has been generated using the updated and more independent method which is presented in this paper.

moving average over 20 patches – approximately 0.2 s – for the distance difference derivative and plotted the result in Fig. 10. The plot can be compared to Fig. 9, where no averaging has been done. We see that the moving average substantially reduces the noise in the estimates.

Even in Fig. 10, the estimates in the beginning are noisy. This is due to the character of the song that was played, where the sound is not persistent until after 5–6 s. In the beginning there are just intermittent drumbeats and silence between these. Then the information is not sufficient to make good estimates. Thus, it is more fair to the algorithm to review the results from 5–6 s and forward.

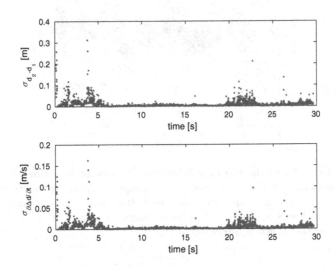

Fig. 11. The standard deviation for parameters that was estimated for the real data. The upper plot shows the standard deviation of the distance difference in Fig. 8 over time and the lower plot shows the standard deviation of the derivative of the distance difference in Fig. 9.

Estimating the Standard Deviation of the Parameters. We have also computed the standard deviations of the parameters in accordance to Sect. 3.1. These are plotted over time in Fig. 11. We can see that the estimations are more uncertain in the beginning of the song, in consistence with when the signal is not persistent. However, just by looking at the estimated Doppler factor this seems to be more uncertain than the theoretical standard deviation suggests.

We also estimated the standard deviations empirically. This was done using the results in Figs. 8 and 9. The empirical standard deviation was computed for the difference between our estimations and ground truth, for a certain time window, namely $t \in [10, 15]$.

The different standard deviations are displayed in Table 2. For the theoretical values we have computed the mean and median, both for all signal and for $t \in [10, 15]$ for comparison with the empirical values.

We can see that the theoretical and empirical values agree quite well for the translation. The reason that the mean of the theoretical standard deviation is higher for all signal is due to the parts of the signal that are more uncertain. However, in the chosen time window the values agree well.

For the Doppler factor the theoretical standard deviation is lower compared to the empirical estimates. This is interesting and there can be several reasons.

Table 2. The mean and the median for the standard deviation of the estimated distance difference (Fig. 8) and the Doppler factor (Fig. 9) for the two received signals.

		Translation, $d_2 - d_1$	Doppler factor, $\partial \Delta d / \partial t$
Theoretical, all signal	Mean of std	$1.03 \cdot 10^{-2}$	$5.25 \cdot 10^{-3}$
	Median of std	$4.71 \cdot 10^{-3}$	$2.39 \cdot 10^{-3}$
Theoretical, $t \in [10, 15]$	Mean of std	$4.58 \cdot 10^{-3}$	$2.29 \cdot 10^{-3}$
	Median of std	$4.12 \cdot 10^{-3}$	$2.08 \cdot 10^{-3}$
Empirical, $t \in [10, 15]$		$3.88 \cdot 10^{-3}$	$4.43 \cdot 10^{-1}$

To begin with, we made some assumptions for the received signals when we derived the equations in Sect. 3.1, which are probably not true for our data. E.g. in our experiments we estimated the noise in the signals as the difference between the two signals after modification. In the bottom plot of Fig. 7 we see that there is still an amplitude difference between the two signals. This means that our estimated noise will not be w.s.s., as was assumed in the derivations. Furthermore, the noise will thus be overestimated. Actually, it turned out the SNR was below 4.7.

Except from this, our method is developed to work with one signal with constant parameters and does not take into account that the patches in our real data actually constitutes one long signal. Also, we might have forgotten to take some important factor into account in out derivations for the standard deviation of the Doppler factor. It might be that the problem cannot be modeled as linear. Regardless, an interesting point for future focus is to investigate this.

4.3   Expanding the Model for Real Data

As mentioned in Sect. 3.2 it is in practice not much harder to estimate three model parameters. Therefore, to get a more precise solution (see Sect. 4.1 and the end of the previous section), we have also made experiments on the same data using (42) as model for the signals. The computations are made in the same manner as in the previous section but the error function (45) is replaced by

$$\int_t (V^{(i)}(t) - \gamma^{(i)} \bar{V}^{(i)}(\alpha^{(i)} t + \tilde{h}^{(i)} + h^{(i)}))^2 \, dt, \tag{55}$$

and the optimization is performed over all three parameters, the subsample translation $h^{(i)}$, the Doppler factor $\alpha^{(i)}$ and the amplitude factor $\gamma^{(i)}$.

The results from using this model for the same signal patch as in Fig. 7 can be seen in Fig. 12. After moving the signals according to the estimated parameters the norm of the difference between the signals (bottom plot in the figures) has decreased with 20% when we included the amplitude factor compared to when we did not.

The plots for the translation parameter and the Doppler factor look similar to the plots in Figs. 8 and 9. However, we can now make a comparison to ground truth for the amplitude factor γ as well.

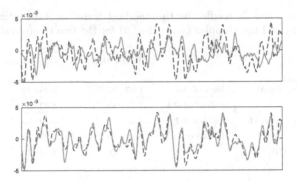

Fig. 12. The plot shows the same signal patches as in Fig. 7. The difference is that a larger model, namely (42), has been used here and thus an amplitude has been estimated as well. The bottom image shows the same signals after modifications using the optimal parameters.

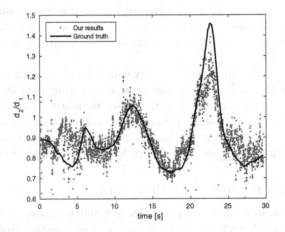

Fig. 13. The distance quotient d_2/d_1 plotted over time. The solid line ( ) represents the ground truth and each dot (•) is the estimation for a certain patch. While the estimations are somewhat noisy there is no doubt that the pattern is the same. The plot is similar to Fig. 9 in [10], but has been generated using the updated and more independent method which is presented in this paper.

The amplitude difference of the two received signals can be compared to $d_1^{(i)}$ and $d_2^{(i)}$. The amplitude estimate $\gamma^{(i)}$ is related to the quotient of the distances, $d_2^{(i)}/d_1^{(i)}$. Since the sound spreads as on the surface of a sphere, the distance quotient is proportional to the square root of the amplitude $\gamma^{(i)}$,

$$\frac{d_2^{(i)}}{d_1^{(i)}} = C \cdot \sqrt{\gamma^{(i)}}. \tag{56}$$

The unknown constant C depends on the gains of the two recording channels. For the experiment, the estimated proportionality constant was $C = 1.3$.

The distance quotient is plotted over time in Fig. 13 – our estimations as dots and ground truth as a solid line. Again we see that they clearly follow the same pattern.

5   Conclusions

In this paper we have studied how to estimate three parameters – time-differences, amplitude changes and minute Doppler effects – from two audio signals. The study also contains a stochastic analysis for these estimated parameters and a comparison between different signal models. The results are important both for simultaneous determination of sender and receiver positions, but also for localization, beam-forming and diarization. In the paper we have built on previous results on stochastic analysis of interpolation and smoothing in order to give explicit formulas for the covariance matrix of the estimated parameters. In the paper it is shown that the approximations that are introduced in the theory are valid as long as the smoothing is at least 0.55 sample points and as long as the signal-to-noise ratio is greater than 4.7. Furthermore, we show using experiments on both simulated and real data that these estimates provide useful information for subsequent analysis.

Acknowledgements. This work is supported by the strategic research projects ELLIIT and eSSENCE, Swedish Foundation for Strategic Research project "Semantic Mapping and Visual Navigation for Smart Robots" (grant no. RIT15-0038) and Wallenberg Autonomous Systems and Software Program (WASP).

References

1. Anguera, X., Bozonnet, S., Evans, N., Fredouille, C., Friedland, G., Vinyals, O.: Speaker diarization: a review of recent research. IEEE Trans. Audio Speech Lang. Process. **20**(2), 356–370 (2012)
2. Anguera, X., Wooters, C., Hernando, J.: Acoustic beamforming for speaker diarization of meetings. IEEE Trans. Audio Speech Lang. Process. **15**(7), 2011–2022 (2007)
3. Åström, K., Heyden, A.: Stochastic analysis of image acquisition, interpolation and scale-space smoothing. In: Advances in Applied Probability, vol. 31, no. 4, pp. 855–894 (1999)
4. Batstone, K., Oskarsson, M., Åström, K.: Robust time-of-arrival self calibration and indoor localization using wi-fi round-trip time measurements. In: Proceedings of International Conference on Communication (2016)
5. Brandstein, M., Adcock, J., Silverman, H.: A closed-form location estimator for use with room environment microphone arrays. IEEE Trans. Speech Audio Process. **5**(1), 45–50 (1997)
6. Cirillo, A., Parisi, R., Uncini, A.: Sound mapping in reverberant rooms by a robust direct method. In: IEEE International Conference on Acoustics, Speech and Signal Processing, pp. 285–288, April 2008

7. Cobos, M., Marti, A., Lopez, J.: A modified SRP-PHAT functional for robust real-time sound source localization with scalable spatial sampling. IEEE Signal Process. Lett. **18**(1), 71–74 (2011)
8. Crocco, M., Del Bue, A., Bustreo, M., Murino, V.: A closed form solution to the microphone position self-calibration problem. In: ICASSP, March 2012
9. Do, H., Silverman, H., Yu, Y.: A real-time SRP-PHAT source location implementation using stochastic region contraction (SRC) on a large-aperture microphone array. In: ICASSP 2007, vol. 1, pp. 121–124, April 2007
10. Flood, G., Heyden, A., Åström, K.: Estimating uncertainty in time-difference and doppler estimates. In: 7th International Conference on Pattern Recognition Applications and Methods (2018)
11. Knapp, C., Carter, G.: The generalized correlation method for estimation of time delay. IEEE Trans. Acoust. Speech Signal Process. **24**(4), 320–327 (1976)
12. Kuang, Y., Åström, K.: Stratified sensor network self-calibration from tdoa measurements. In: EUSIPCO (2013)
13. Kuang, Y., Burgess, S., Torstensson, A., Åström, K.: A complete characterization and solution to the microphone position self-calibration problem. In: ICASSP (2013)
14. Kuang, Y., Åström, K.: Stratified sensor network self-calibration from tdoa measurements. In: 21st European Signal Processing Conference 2013 (2013)
15. Lindeberg, T.: Scale-space theory: a basic tool for analyzing structures at different scales. J. Appl. Stat. **21**(1–2), 225–270 (1994)
16. Lindgren, G., Rootzén, H., Sandsten, M.: Stationary Stochastic Processes for Scientists and Engineers. CRC Press, New York (2013)
17. Petersen, K.B., Pedersen, M.S., et al.: The matrix cookbook. Tech. Univ. Den. **7**(15), 510 (2008)
18. Plinge, A., Jacob, F., Haeb-Umbach, R., Fink, G.A.: Acoustic microphone geometry calibration: an overview and experimental evaluation of state-of-the-art algorithms. IEEE Signal Process. Mag. **33**(4), 14–29 (2016)
19. Pollefeys, M., Nister, D.: Direct computation of sound and microphone locations from time-difference-of-arrival data. In: Proceedings of ICASSP (2008)
20. Shannon, C.E.: Communication in the presence of noise. Proc. IRE **37**(1), 10–21 (1949)
21. Zhayida, S., Segerblom Rex, S., Kuang, Y., Andersson, F., Åström, K.: An Automatic System for Acoustic Microphone Geometry Calibration based on Minimal Solvers. ArXiv e-prints, October 2016
22. Zhayida, S., Andersson, F., Kuang, Y., Åström, K.: An automatic system for microphone self-localization using ambient sound. In: 22st European Signal Processing Conference (2014)
23. Zhayida, S., Åström, K.: Time difference estimation with sub-sample interpolation. J. Signal Process. **20**(6), 275–282 (2016)

Applications

Detection and Classification of Faulty Weft Threads Using Both Feature-Based and Deep Convolutional Machine Learning Methods

Marcin Kopaczka[1]($\boxtimes$), Marco Saggiomo[2], Moritz Güttler[1], Kevin Kielholz[1], and Dorit Merhof[1]

[1] Institute of Imaging and Computer Vision, RWTH Aachen University, Aachen, Germany
marcin.kopaczka@lfb.rwth-aachen.de
[2] Institut für Textiltechnik, RWTH Aachen University, Aachen, Germany

Abstract. In our work, we analyze how faulty weft threads in air-jet weaving machines can be detected using image processing methods. To this end, we design and construct a multi-camera array for automated acquisition of images of relevant machine areas. These images are subsequently fed into a multi-stage image processing pipeline that allows defect detection using a set of different preprocessing and classification methods. Classification is performed using both image descriptors combined with feature-based machine learning algorithms and deep learning techniques implementing fully convolutional neural networks. To analyze the capabilities of our solution, system performance is thoroughly evaluated under realistic production settings. We show that both approaches show excellent detection rates and that by utilizing semantic segmentation acquired from a fully convolutional network we are not only able to detect defects reliably but also classify defects into different subtypes, allowing more refined strategies for defect removal.

1 Introduction

In modern industrial textile production, automated weaving is one of the key technologies regarding both turnover and total amount of produced fabric. In weaving, two sets of yarns - called weft and warp yarns - are combined on a weaving machine to form a fabric. The weft yarns are a set of usually hundreds of horizontal parallel yarns oriented in production direction that are systematically displaced in vertical direction by a mechanical frame. To form the most basic weaving pattern, the plain weave, every second warp yarn is displaced upwards while the remaining yarns are moved downwards, forming a space between the upward and downward displaced yarns. This space - the shed - allows inserting a further horizontal yarn perpendicular to the weft yarns, and weaving a plain weave itself is the process of adding additional warp yarns repeatedly while switching the upwards and downwards displaced yarns between each insertion.

© Springer Nature Switzerland AG 2019
M. De Marsico et al. (Eds.): ICPRAM 2018, LNCS 11351, pp. 141–163, 2019.
https://doi.org/10.1007/978-3-030-05499-1_8

Several types of weaving machines have been developed and are currently used in textile production. Of these machines, air-jet weaving machines are widely used due to their superior productivity of up to 1500 insertions per minute and their high reliability. In air-jet weaving, compressed air is used for inserting the warp yarn into the shed. Additional air ducts maintain a defined air flow along the whole shed width, allowing the weft yarn to move quickly between the warp yarns. Thanks to this feature, air-jet weaving machines offer the fastest production speeds in terms of yarn insertions per minute. However, in some cases the weft yarn may not reach its destination as it collides with a warp yarn or it leaves its desired trajectory for other reasons such as inhomogeneities in the air flow. Weaving these faulty weft threads into the textile would alter the structure of the fabric, resulting in defects that negatively affect the appearance and mechanical properties of the fabric. To avoid this problem, air-jet weaving machines are equipped with a sensor that detects whether the weft yarn has been inserted correctly into the shed. More precisely, weft yarn defects are detected with photoelectric or laser-based weft sensors such as [1] that detect whether the weft yarn arrived correctly at the receiving end of the machine. Modern automated weaving machines are also equipped with an automated defect removal system that inserts an additional thread of yarn into the shed while the machine is stopped. The second thread collides with the faulty thread in the shed, and momentum causes both threads to move towards the receiving side of the machine where a tube generating negative pressure removes both yarns, leaving a clean shed.

Currently, machine vision is not applied to monitor machine or shed status. However, quality inspection systems based on computer vision for defect detection in textiles are an active area of research. The reasons behind fabric inspection being such an active field are two-fold: First, manual textile inspection is an expensive and time-consuming task as it requires direct screening performed by qualified personnel. At the same time, experiments have shown that human performance in textile inspection degrades quickly and that humans find no more than 60% of defects in textile fabric [2]. This emphasizes the need for novel approaches allowing increasing defect detection rates and decreasing quality control expenses. The second reason is that textile inspection is a general problem that can be solved using different algorithms originating from various fields of computer vision. Therefore, a broad range of algorithms relying on a large number of underlying algorithms has been developed and adapted to allow localizing and identifying defects in textiles. Notable reviews include [3], where also an overall taxonomy of fabric inspection methods is introduced, furthermore in [4], where the authors extend the range of algorithms to include methods that allow defect detection in patterned fabrics, and most recently in the work by [5], which also puts emphasis on the image acquisition and imaging hardware.

While the devices used for weft yarn monitoring - including the laser sensors described above - can be defined as optical sensors, their output signal is not an image but a purely scalar value that does not allow applying image processing algorithms. modern machines are equipped with an automated defect removal mechanism that allows ejecting faulty weft threads by inserting an additional yarn into the stopped machine. So, if a faulty weft yarn is detected, then textile production is stopped and the automated weft yarn removal procedure is initiated. However, while the error detection itself is highly reliable, the faulty yarn removal succeeds only in 80% of the cases. In the remaining 20% of the cases, the removal system either fails to remove the yarn or may even increase the complexity of the defect, requiring additional manual effort to clean the shed and allow production to continue. To increase system reliability, we therefore introduce a camera-based inspection system that checks if a faulty weft yarn is present in the shed after a weft yarn defect has been detected by the built-in defect detection system. A combination of specialized imaging hardware and an image processing pipeline developed especially for this task is presented. We analyze the effect of using different methods for image preprocessing that allow enhancing the visual appearance of faulty weft threads in combination with image descriptors sensitive to structures which can be found in our image data. Additionally, we introduce a deep learning pipeline that allows not only faulty weft thread detection on single-pixel level, but additionally enables classification into automatically removable defects and those requiring manual removal by an operator, thereby saving time when a defect can be handled by the machine itself and reducing risk of additional issues that may arise when the machine attempts removing a complex defect automatically. The system is described in detail and its performance on a real industrial weaving machine is evaluated throughly, allowing to define the optimal pipeline to solve the problem of faulty weft yarn detection. To the best of our knowledge, our approach is the first to allow highly reliable weft yarn defect detection and defect classification using computer vision.

2   Previous Work

In recent research publications, a number of algorithms for vision-based defect detection has been proposed. Of the available approaches, a specific subgroup of methods related closely to our research area are systems for online (or, in our specific context, onloom) analysis of woven fabrics using a camera system attached directly to the loom. Working directly in line with the textile creation requires highly efficient algorithms, as the automated defect detection needs to meet hard real-time requirements to meet the production speeds of modern weaving machines. An early approach has been presented almost two decades ago by [7], where yarn density is measured on the loom by applying Fourier transform to images acquired by a camera attached to the machine. The authors of [8] introduce a system for onloom defect detection in woven fabrics and subsequently test their approach on a loom mockup. In their algorithm, defects such as holes in the textile are detected using texture analysis methods. The currently most advanced system has been introduced in [9], where a vibration-stabilized movable high resolution camera that traverses perpendicular to the production direction is attached to the weaving machine, allowing capturing images with a spatial resolution of 1000 dpi in real-time during production. Performance-optimized

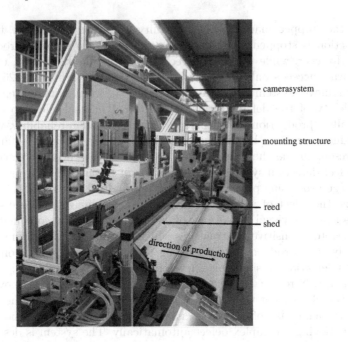

camerasystem

mounting structure

reed

shed

direction of production

Fig. 1. The visual inspection system mounted to an automatic weaving machine. Previously published in [6].

and image processing algorithms with GPU support for faster computation allow the precise localization of single yarns in the produced fabric and thereby the detection of various types of challenging defects in real-time.

In contrast to previously published methods, our method does not analyze the produced fabric after defects may have occured, but instead it focuses on the detection of faulty weft yarns in the shed. The results of our algorithms allow the operator to react and correct the machine state before any defective fabric is produced. Furthermore, our algorithms can decide whether an automated removal is possible or if an operator needs to be informed to perform manual defect correction. This approach is highly innovative as it allows to increase product quality by avoiding defects caused by weaving defective yarns into the product. Parts of this work have been previously published in [6]. This article extends the original publication by introducing additional methods for defect detection using deep learning techniques and feature descriptors that can take advantage of the detailed output provided by the deep neural networks. The novel methods are motivated, described in detail and evaluated on an additional dataset that has been annotated by experts to allow not only the detection of detects, but also their classification into defects that are automatically removable using the machine's built-in systems and those that require operator interaction.

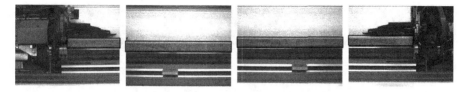

Fig. 2. Top view of the shed as captured by the four cameras. The used ROIs are outlined in red. Previously published in [6]. (Color figure online)

3 Materials and Methods

This section gives detailed information on the developed methods. For the classical machine learning approach, we start with a description of the image acquisition setup and basic data preprocessing steps, followed by our processing pipeline for image enhancement, feature extraction and classification. Finally, we describe the decision rule used to compute the confidence if an alert should be raised or not. In the part describing the semantic segmentation using a fully convolutional neural network (FCN), we describe the network architecture followed by a set of specialized descriptors for defect classification.

3.1 Overview

Figure 3 shows our feature-based pipeline from a high-level perspective. A multi-camera array is used to acquire shed images that are subsequently processed in order to allow feature extraction and classification of small image patches. Based on the patch-based classification, a final per-image classification is implemented to automatically decide if an image contains a faulty weft yarn. The following subsections provide a more detailed description of the implemented steps. The approach based on convolutional neural networks is described in Sect. 3.5

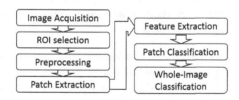

Fig. 3. Overview of the implemented pipeline. Previously published in [6].

3.2 Camera Setup and Image Preparation

Image acquisition was performed using an array of four industrial cameras. The cameras were spaced equidistantly in a sensor bar above the shed with the cameras facing downward, allowing a clear view at the shed's contents at a resolution

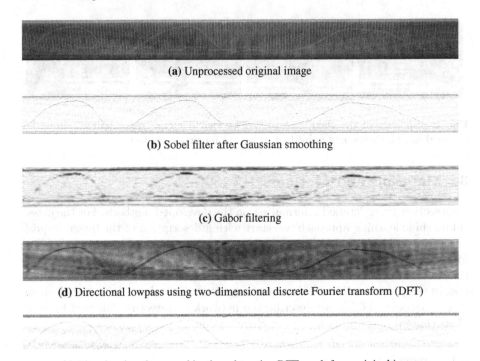

(a) Unprocessed original image

(b) Sobel filter after Gaussian smoothing

(c) Gabor filtering

(d) Directional lowpass using two-dimensional discrete Fourier transform (DFT)

(e) Directional unsharp masking by subtracting DFT result from original image

Fig. 4. Visual appearance of the different preprocessing methods applied to the images in order to enhance defect visibility. Preprocessing results are shown with inverted gray values to increase contrast for the reader. Previously published in [6].

of 200 dots per inch (dpi) (Fig. 1). Several options such as a single-camera setup with a traversing camera system or a single camera with a sufficiently large field-of-view are also possible options to acquire images at a sufficient resolution, however implementing an array of several cameras instead of a single fixed or moving camera yields several benefits:

- A multi-camera setup allows fast adaptation to wider or smaller machines by chosing the number of cameras accordingly.
- The shed has a strongly elongated shape, with a width of about 2 m depending on the machine and a depth of about 10 cm. A multi-camera setup with a wide combined field of view allows to cover the shed area more efficiently than a single camera with 4:3 aspect ratio.
- Using several cameras allows picking devices with a lower spatial resolution than a single camera that covers the whole shed area while maintaining a high image resolution.
- While the same benefits could have been gained by using a single moving camera as in [9], constructing a traversing system poses additional challenges such as camera position retrieval and potentially motion blur that can be

avoided by using an array of cameras mounted in fixed positions. At the same time, the additional cost for using multiple cameras is leveraged by the fact that a moving solution requires additional precise motors and possibly a damping system that also contribute to the system's total cost.

The area captured by the cameras was still larger than the shed itself, therefore fixed regions of interest (ROI) were defined for each of the cameras in which all regions except the shed were masked out. All subsequent image analysis was performed exclusively on the ROIs, ignoring the non-shed parts of the image. Figure 2 shows full images and the respective ROIs used for further processing.

3.3   Preprocessing

Several preprocessing steps aimed at enhancing defect appearance have been implemented in order to increase the appearance of relevant image structures. Since the dominant structure in defect-free images is the repeating, vertical pattern formed by the weft yarns and defects break this regularity by introducing non-vertical components, the chosen preprocessing steps aim at increasing the contrast between defect and non-defect areas by enhancing defect appearance. To this end, we take advantage of the direction difference and use orientation-sensitive preprocessing methods. A sample image and filter results computed from it are shown in Fig. 4. The filters include:

- **Basic edge and contour detectors**; we picked the Sobel and Prewitt gradient operators. These filter kernels allow a good enhancement of horizontal structures when applied in their horizontal form while at the same time being computationally highly efficient. Both are known to be sensitive to noise and grain in the image data, therefore we also analyzed the impact of adding an additional Gaussian smoothing filter before computing the gradients. A sample result is shown in Fig. 4b.
- **Gabor Filters**, a group of algorithmically defined filter kernels that allow enhancing image structures with a defined orientation and length [10]. In comparison to the basic Sobel and Prewitt filters, the multiparametric nature of Gabor filters allows more fine-grained control over the type of structure to be enhanced. While gabor filters are commonly used in groups of differently parametrized kernels as filter banks for texture and image classification, in our preprocessing implementation we use only horizontal Gabor filters with a constant length to enhance horizontal structures as shown in Fig. 4c.
- **Spectral lowpass filters** based on Discrete Fourier Transform (DFT). They allow controlling filter characteristics in a very fine-grained manner. In more detail, the filter is applied by computing the DFT of the image, centering the resulting spectrum and multiplying it with a binary mask. This approach allows eliminating vertical structures in the image while enhancing horizontal patterns at the same time. To obtain the final result, we apply inverse DFT to the computed lowpass-filtered spectrum (Fig. 4d). This approach can be extended by subtracting the lowpass result from the original image, leaving

only high-frequency components such as the yarns. When analyzing the shed images, this solution eliminates the regular pattern formed by the warp yarns and thereby allows further enhancement of the relevant weft yarns (Fig. 4e).

3.4   Feature Extraction and Image Classification

We perform patch-wise classification of the shed images by splitting the full images into smaller patches after applying one of the preprocessing steps described above. Subsequently, classification is performed by computing feature or texture descriptors and feeding the resulting feature vectors into a machine learning-based classifier. The feature extractors and classifiers used in this work will be described in this section. A method for labeling each of the patches differently depending on whether or not they show defective warp threads has been developed as well, since the classifiers applied are supervised learning methods and thereby require precise image labels. This section gives a detailed overview of all steps involved in the classification chain.

First, we extract small non-overlapping quadratic patches sized 30×30 pixels from the ROIs. This value for patch size has been determined in preliminary experiments. Working on patches instead of whole images yields several advantages: First, it allows to drastically increase the amount of labelled training images for the classifiers and reduces feature vector length at the same time. Both of these properties are factors for successful training of classifiers, allowing greater reliability and faster classification speed. Additionally, classifying small image fragments with well-known locations within the image allows a precise localization of defect warp yarns, thereby allowing analysing defect positions for further processing. Also, full images of defect- and non-defect cases are often highly similar except for the image areas covered by the faulty yarn; therefore using small patches increases the inter-class difference between defect and non-defect patches. After extracting the patches, we assign a label to each patch depending on the amount of defective thread visible in the center area of the patch. A well-performing decision rule found by evaluating preliminary experiments is declaring a patch defective if it contains at least 50 defective pixels.

After labeling the acquired image patches, we perform feature extraction to allow better descriptions of the image data. From the available methods, we have picked feature descriptors that either have been proven to perform well in textile image analysis tasks or have properties that allow enhancing visual structures that are likely to occur in our data:

- The **pixel intensity values** themselves. Feeding the image directly into the classifier is computationally highly efficient as it requires no additional computing time for feature extraction and allows to analyze absolute intensities and local structures (Fig. 5a). However, this approach has the downside of not providing gradient and neighborhood information that might allow improved classifier performance. Furthermore, this method is more sensitive to shift and rotation than some descriptor-based methods.

- **Histograms of Oriented Gradients (HOG)**, a widely used image descriptor that creates feature vectors by extracting local gradient information at the cost of discarding the intensity data [11]. As Fig. 5b shows, a well-parametrized HOG operator results in feature vectors that discriminate well between defect and non-defect patches by detecting horizontal edges that in our case appear very often in defect cases and less often in non-defect patches. A detailed analysis of the performance of different HOG variants and parameter sets is given in Sect. 4.
- **Local Binary Patterns (LBPs)**, another well understood method that extracts statistical neighborhood information to form a texture descriptor [12]. In our work, we have decided to use the homogeneous LBP variant, as it has shown better performance for our problem than its rotation-sensitive counterpart (See Sect. 4).

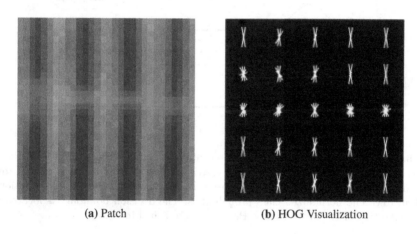

(a) Patch   (b) HOG Visualization

Fig. 5. Feature extraction is done on patches with 30×30 pixels. Previously published in [6].

The extracted feature vectors were subsequently fed into a binary classifier (defect vs. non-defect) that allowed computing label predictions for each patch. In this work, we tested the performance of several widely used state-of-the art classifiers: The k-nearest-neighbors classifier (kNN), binary decision trees (BDT) and random forests (RF) [13]. For our evaluation, the key parameters of the classifiers were varied systematically and optimal values for our problem were defined (Sect. 4). The final decision whether a defect is visible in the image is computed using a maximum a posteriori (MAP) approach that takes into account that the classes may be imbalanced, i.e. the probability of having a defect or a non-defect image is not equal (see [14] for details on MAP). This method computes class probabilities by multiplying a likelihood with a prior, with the likelihood being the observation - in our case the percentage of patches in an image reported as defective - and the prior is the known base probability of an image displaying a defect, i.e. the ratio of defect vs. non-defect images in the training set. Figure 6 shows the process in more detail.

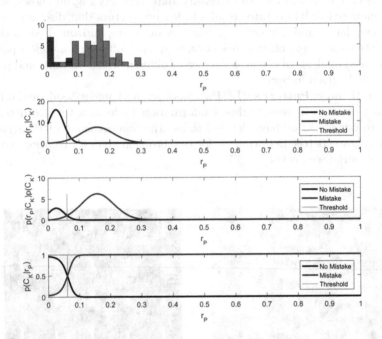

Fig. 6. MAP procedure. Top row: The percentage of patches classified as defective is noted in a histogram. 2nd row: Gaussian distributions are fitted to the data. 3rd row: Both distributions are multiplied with a-priori knowledge about the class distributions. Bottom row: The normalized class probabilities. The decision boundary is located at the intersection of the probabilities. Previously published in [6].

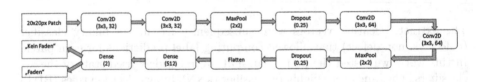

Fig. 7. Our deep convolutional network architecture.

3.5   Precise Defect Detection Using a Fully Convolutional Network

We apply recent advances in deep learning to implement an alternative app-roach to the defect detection problem. The annotated images are fed into a fully connected convolutional neural network for semantic segmentation, returning a pixel-wise probability map from which the exact position of the faulty weft yarn is derived subsequently. This approach returns more detailed information on the defect morphology, allowing not only reliable defect detection but also a subsequent categorization of different defect subclasses. In our case, we grouped all possible defects into two main classes - those that can be removed reliably by the machine's built-in defect removal system and those that are too complex for automated removal and require a trained machine operator to remove the thread manually since using the automated system in this case would most probably fail and might potentially result in even larger defects. Figure 9 shows an overview of our deep classification pipeline.

In a first step, all ROIs covering the shed are normalized using CLAHE equalization to even out illumination variation and subsequently merged to obtain an image of the entire shed area. The resulting images are fed into convolutional networks designed for semantic segmentation. We implemented two networks for this task and evaluated their performance - a lightweight, custom convolutional network for pixel classification (See Fig. 7 and the U-Net [15], a well-established state-of-the-art FCN. The exact U-Net architecture used in our work is shown in Fig. 8. The custom net returns a binary value of 0 for no thread and 1 for a detected thread pixel, while the U-Net computes a continuous value between 0.0 (guaranteed to contain no faulty thread) and 1.0 (guaranteed to contain a faulty thread) for each pixel. The values are subsequently multiplied by 255 to allow efficient mapping to 8-bit values and storing them as regular grayscale images.

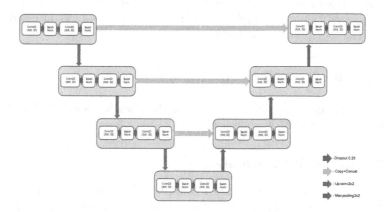

Fig. 8. Our U-Net implementation.

After segmentation was performed, the segmentation masks were fed into feature extractors. We used two efficient feature extractors for defect classification:

- The **weight distribution feature** computes the sum of all pixels in a given area of the segmentation mask. Since pixels where a faulty weft thread has been located by the FCN are given a high value, this value is high for areas where faulty weft threads are located. Complex defects resulting in loops show an even higher pixel count, making this feature suitable for both defect detection as well as distinguishing between automatically removable defects and those requiring manual intervention.

- The **color change feature** is applied to a single vertical column of the segmentation image and counts how often the probability along the column flips between below and above 0.5 (or 127 in the case of 8-bit images) and vice versa. If no yarn is in the column, then the number of flips is zero. If a single yarn is detected, then the number of flips is two (one from a low probability value to a high one when the yarn pixels start and one at their end). Therefore, this feature allows assessing the complexity of a defect since its value increases drastically when complex patterns such as loops are crossed by the column.

We divided the full shed image horizontally into twelve equally-sized segments and computed the weight distribution feature for each segment, while the color change feature was computed for each segment's first column. Subsequently, the data gathered by the descriptors has been aggregated in two different ways: Spatial ordering of each feature, where each segment has a fixed place in the feature vector, and histogram ordering, where the features were organized in a 12-bin histogram. The first method allows identifying the exact locations of areas with higher pixel count, while using the histogram allows a fast statistical analysis of the entire shed. The resulting four combinations of two feature descriptors are fed into a random forest classifier that is evaluated in the results section of this chapter. Figure 10 shows a sample image and the output and weight distribution of each of the four possible methods.

4   Experiments and Results

Here, we evaluate the performance of both the classification based on feature descriptors and of the FCN approach.

4.1   Feature-Based Classification

The experiments for evaluating the feature-based methods served two major goals: Evaluation and optimization of classifier performance and validation of the developed algorithms on realistic data. To this end, the system was mounted on a modern industrial automatic weaving machine (Dornier A1) and 75 sets of four images each were acquired; 60 of the sets were showing a faulty warp thread and the other 15 were defect-free. We used precision and recall for performance evaluation, defined as

$$\text{precision} = \frac{\text{number of correctly detected defects}}{\text{total number of detections}}$$

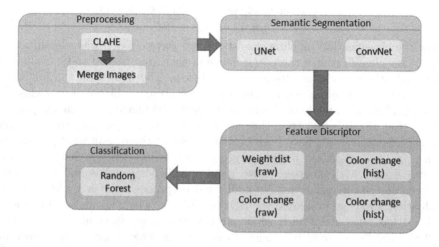

Fig. 9. Deep defect classification: Images from all four cameras are merged and preprocessed using CLAHE. The UNet and a custom fully convolutional network are used to segment the images. A set of custom-designed feature descriptors (see text for details) is applied for feature extraction. Subsequently, a random forest classifier is applied to the feature vector for classifying the images into defect/non-defect images and subclassifying the defects into those that require manual action and those that can be resolved automatically using the machine's automated defect removal system.

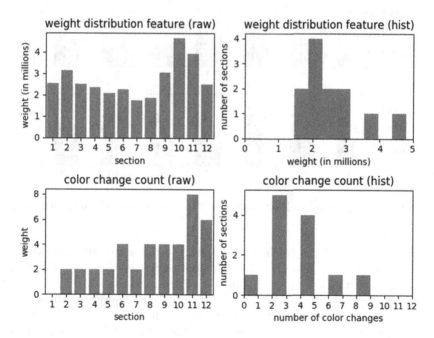

Fig. 10. Sample image and computed features returned by the four presented descriptors.

$$\text{recall} = \frac{\text{number of correctly detected defects}}{\text{total number of defects}}.$$

Since the classification pipeline consists of two steps - patch-wise classification and image classification - the performance of each step needed to be evaluated independently and the results are shown below. The figures on the following pages each show two bar charts. The upper one presents the patch-wise classification performance, while the lower one presents the image classification performance (Fig. 13). In a first trial, the impact of descriptor type and parameters on the classification result have been evaluated. All of these tests were performed using the same classifier; preliminary experiments have shown that a random forest classifier with 30 trees, shows good overall performance and robustness towards descriptor variations. No additional preprocessing was applied for the evaluation. First, we evaluated the HOG descriptor, where a constant block size of 1×1 has been maintained while varying histogram bin count and cell size. Cell size has been varied in a range from 3×3 to 30×30 pixels. As shown in Fig. 11, classifier performance remains stable over a wide range of settings with only very small and very large cell sizes having a negative impact on the result. The impact of the histogram's bin count at a constant cell size of 6×6 pixels is shown in Fig. 12. Overall, the HOG descriptor shows stable rates with a small tendency towards higher performance with increasing with increasing bin count.

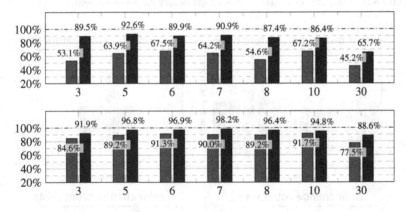

Fig. 11. HOG: Quadratic cell size variation for 7 bins. Previously published in [6].

For LBPs, we analyzed the impact of using rotation-invariant versus conventional LBPs, the effect of varying the number of neighbors, the effect of changing the descriptor radius and finally the impact of manipulating cell size. As shown in Figs. 14, 15, 16 and 17, varying rotational variance or descriptor radius have a strong impact on the classification performance, while the effects of varying cell size or the neighbor count are more subtle. Unless otherwise specified, cell size is 6×6, neighbor count is 7, radius is 1 and the rotationally invariant version is used.

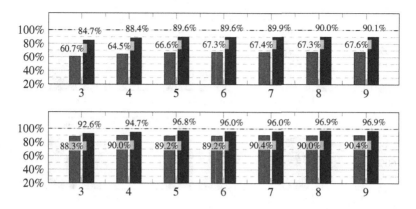

Fig. 12. HOG: Histogram bin count variation for 6×6 cells. Previously published in [6].

Fig. 13. Legend for bar charts. The upper charts of the figures present the patch-wise classification performance. The lower charts present the image classification performance.

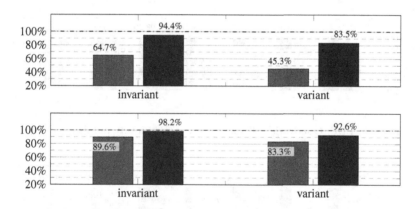

Fig. 14. LBP: Rotational variance. Previously published in [6].

After analyzing descriptors and choosing optimized parameter settings, we used HOG or LBP results as feature vectors and evaluated the effects of the different implemented preprocessing methods when classifying using either pixel intensity values or preprocessed images. The results are shown in Figs. 18, 19 and 20. It can be seen that if no feature descriptor is used, then preprocessing has the highest impact on classification outcome. Gabor and Sobel-of-Gaussian filtering achieve the best classification rates with a result of 96.4%. Using HOG as feature descriptor lowers the effect of image-enhancement using preprocessing,

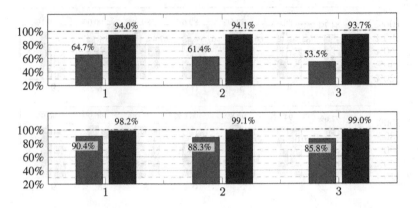

Fig. 15. LBP: Radius variation for cell size 7 × 7. Previously published in [6].

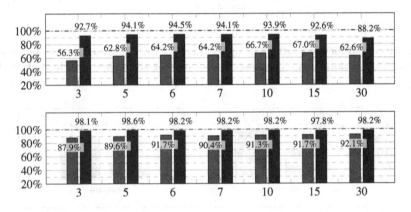

Fig. 16. LBP: Quadratic cell size variation. Previously published in [6].

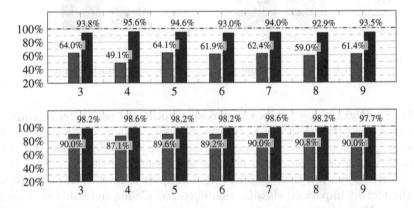

Fig. 17. LBP: Neighbor count variation. Previously published in [6].

indicating that this descriptor is able to extract its features optimally from original unprocessed images. When using HOG, using no preprocessing outperforms the other methods with a whole-image classification rate of 96.1%. When Gabor filtering is applied to the input images and HOG is computed on the Gabor results, whole-image classification results are good while at the same time showing slightly lower patch-wise classification performance. Finally, LBPs show a constantly high classification rate regardless of the used feature vectors, making them robust to preprocessing algorithms and peaking around 90% without preprocessing as well as for the Sobel and Prewitt filters.

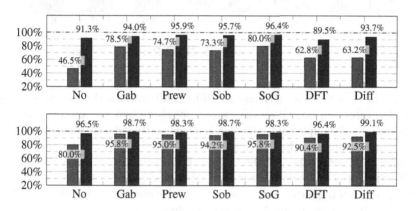

Fig. 18. Different preprocessing methods for pixel intensity feature vector. Previously published in [6].

We have chosen three of the best performing preprocessing/feature descriptor combinations and analyzed how the classification rates vary when different classifiers are used and also inspected how changing the key parameters of the clas-

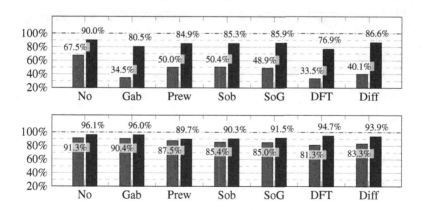

Fig. 19. Different preprocessing methods for HOG. Previously published in [6].

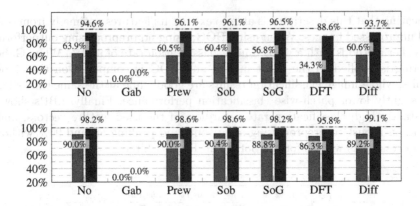

Fig. 20. Different preprocessing methods for LBP. Previously published in [6].

sifiers affect classification performance. The combinations that we analyze here are: Gabor filter and pixel intensity values, Sobel-of-Gaussian and pixel intensity values, and finally applying LBPs to a non-preprocessed image (Fig. 24). These are the three pipelines that have shown the best image classification rates in the previous experiments described above. For this stage, we analyzed the patch classification accuracy, assuming that an increased patch-wise classification rate corresponds to a higher classifier robustness, thereby increasing chances of correctly classifying new, more challenging patches.

The first inspected classifier is kNN, with its only adjustable parameter being the number of neighbors k. The results in Fig. 21 shows that the choice of k has only marginal impact on classification rates. The classifier performance is constantly high even with only one contributing neighbor, while the recall drops continuously with increasing neighbor count.

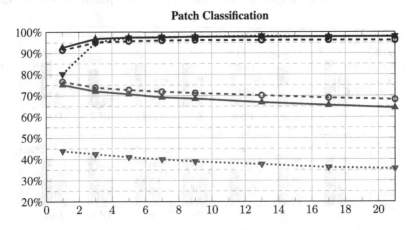

Fig. 21. KNN: Variation of k. Previously published in [6].

Patch Classification

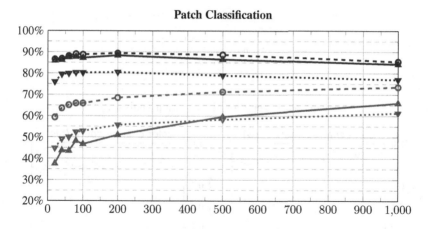

Fig. 22. BDT: Variation of depth. Previously published in [6].

Patch Classification

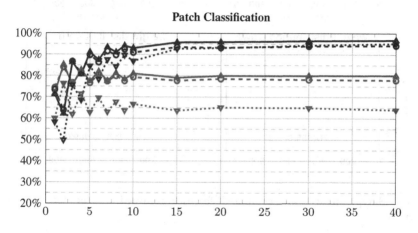

Fig. 23. RF: Variation of number of trees. Previously published in [6].

An analysis of the binary decision tree (Fig. 22) shows two effects: Trees with lower depth yield higher precision values and increasing the number of tree nodes increases recall for all descriptors. In a subsequent step we also changed the number of trees in the random forest classifier while leaving each tree's depth at its default value (Fig. 23). Results show that classification results increase with the number of trees. This effect continues until a number of about 15 trees is reached; from that point on the classification rates remain constant.

Finally, we used the optimized per-patch results to define a decision rule allowing making the final decision whether a ROI contains a defective thread or not. We analyzed the number of positives returned by the classifiers (corresponding to the likelihood of a defect being present in the image or not) for both defect and non-defect ROIs. Additionally, the values were normalized by image

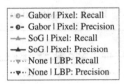

Fig. 24. Legend for classification analysis.

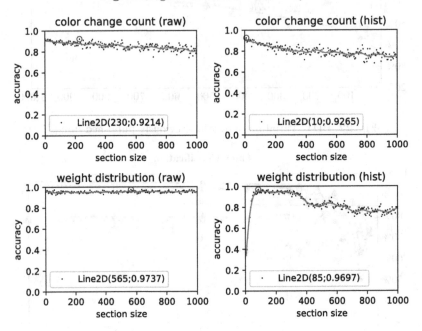

Fig. 25. Sample image and computed features returned by the four presented descriptors.

size. Our data shows that defect and non-defect images form two distinctive distributions in terms of reported defect patches. In order to find the optimal decision rule, we multiplied the likelihoods with the prior class distributions computed from the training image set. Per-image classification rates are displayed in the bottom bar charts of image Figs. 11, 12, 13, 14, 15, 16, 17, 18, 19 and 20. They indicate that the MAP approach has a high robustness and allows classification of images with high accuracy even when results from classifiers with strongly varying patch-wise classification performance are used as input.

4.2   Performance of the Deep Learning Approach

For evaluating the deep approaches, 50 of the 75 sets (200 sub-images in total) were randomly sampled from the database and the remaining 25 sets (100 images in total) were used for testing. The ConvNet was fed and evaluated with 20×20

pixel patches while the U-net was fed with the full images and evaluated on a per-pixel basis, hence the larger number of samples in the evaluation table shown in Tables 1 and 2. The U-Net shows a slightly larger overall accuracy while at the same time displaying a tendency of underestimating the total number of defect pixels. However, result inspection has shown that misclassified pixels usually appear at the border pixels of the thread. This means that while thread thickness is underestimated by the U-Net, it still allows reliable detection of thread position in the shed.

Table 1. ConvNet performance. Overall accuracy: 0.9662.

Custom ConvNet			
	True non-defect	True defect	Precision
Predicted non-defect	105,592	2,520	0.9767
Predicted defect	5,924	135,608	0.9581
Recall	0.9469	0.9818	

Table 2. U-Net performance. Overall accuracy: 0.9662.

U-Net			
	True non-defect	True defect	Precision
Predicted non-defect	32,371,325	435,154	0.9867
Predicted defect	55,358	458,163	0.8921
Recall	1,9982	0,5128	

Table 3. Three-class classification for defect categorization. AR - automatically removable, MR - manual removal needed.

Defect categorization			
	Expert labels		
	Non-defect	AR	MR
Predicted non-defect	68	0	00
Predicted AR	0	89	2
Predicted MR	0	5	66

After performing segmentation, the four presented feature descriptor combinations have been evaluated with subject to the width of the analyzed shed segment. As Fig. 25 shows, the raw weight distribution feature shows best overall performance while at the same time being most insensitive towards section width changes, indicating that it is the most robust of the presented features.

Optimal section size was identified at a width of 565 pixels, where it returned an accuracy of 97.4%. The classifier used here was a random forest with 33 trees, the optimal number of estimators in the forest was determined using a grid search experiment.

The described classifier was subsequently extended from a two-class classification between defect and non-defect samples to a three-class classification allowing distinguishing between automatically removable defects and those that require human intervention. To obtain true labels, an additional set of 230 full shed images (920 subimages) was acquired on the machine under realistic production conditions. The images were labeled by an experienced machine operator into being defect-free (68 images), showing an automatically removable defect (94 images) or a defect that requires stopping the machine and manually removing the defect (68 images). The 230 images were used to evaluate the classifier in a 10-fold cross validation. As Table 3 shows, the system was able to clearly distinguish between defect and non-defect in all cases. The categorization between automatically and manually removable defects succeeded in 157 of the 162 cases (95.7%).

5   Discussion

Our results show that the problem of localizing defective yarns is well solvable using different computer vision methods. At the same time, we have shown that the different algorithm combinations evaluated in this work differ in their performance. Preprocessing has a high impact on classifier performance and basic edge-enhancing methods allow the best class separability for our problem. These filters outperform direction-sensitive and multistage approaches such as the Sobel-of-Gaussian or Gabor Filters. Per-image classification rate generally remains constantly high regardless of the used preprocessing-feature combination and is mostly robust towards drops in the patch-wise classification performance, indicating that the problem is well suited for machine learning algorithms. Using a convolutional network for defect detection also allows a reliable classification between defect and non-defect images and furthermore allows distinguishing between two different defect types with high accuracy.

6   Conclusion

In this paper, we have described and thoroughly evaluated a pipeline for the fully automated detection of faulty weft threads in automated weaving machines. Our tools include image (pre)processing algorithms for image enhancement, a set of feature descriptors and finally different classification algorithms. In a first step, we developed a robust multi-camera setup that is well suited for use with different machine widths and at the same time capable of capturing the entire shed area with high resolution. The resulting images were either divided into small patches and each patch was classified independently by feature-based methods or fed into a convolutional neural network in full resolution. We described and evaluated a number of methods for preprocessing, feature extraction and feature-based classification as well as for defect detection with deep neural networks. Parameter variations of the extraction and classification methods were analyzed and optimized settings were reported. Furthermore, we introduced a maximum-a-posteriori-based method for the final defect classification of a whole image based on the detection distributions in positive and negative test images. Finally, two different architectures of fully convolutional networks and a set of specialized descriptors were described. Results have shown that our presented methods allow a reliable error detection with high classification rates, making the system suitable for future industrial applications.

References

1. Wada, Y.: Optical weft sensor for a loom (1984) US Patent 4,471,816
2. Karayiannis, Y.A., et al.: Defect detection and classification on web textile fabric using multiresolution decomposition and neural networks. In: The 6th IEEE International Conference on Electronics, Circuits and Systems, Proceedings of ICECS 1999, vol. 2, pp. 765–768. IEEE (1999)
3. Kumar, A.: Computer-vision-based fabric defect detection: a survey. IEEE Trans. Industr. Electron. **55**, 348–363 (2008)
4. Ngan, H.Y., Pang, G.K., Yung, N.H.: Automated fabric defect detection-a review. Image Vis. Comput. **29**, 442–458 (2011)
5. Hanbay, K., Talu, M.F., Özgüven, Ö.F.: Fabric defect detection systems and methods-a systematic literature review. Optik Int. J. Light Electron. Opt. **127**, 11960–11973 (2016)
6. Kopaczka, M., Saggiomo, M., Guettler, M., Gries, T., Merhof, D.: Fully automatic faulty weft thread detection using a camera system and feature-based pattern recognition. In: Proceedings of the 7th International Conference on Pattern Recognition Applications and Methods (ICPRAM), pp. 124–132 (2018)
7. Sari-Sarraf, H., Goddard, J.S.: Vision system for on-loom fabric inspection. IEEE Trans. Ind. Appl. **35**, 1252–1259 (1999)
8. Stojanovic, R., Mitropulos, P., Koulamas, C., Karayiannis, Y., Koubias, S., Papadopoulos, G.: Real-time vision-based system for textile fabric inspection. Real Time Imaging **7**, 507–518 (2001)
9. Schneider, D., Holtermann, T., Merhof, D.: A traverse inspection system for high precision visual on-loom fabric defect detection. Mach. Vis. Appl. **25**, 1–15 (2014)
10. Bovik, A.C., Clark, M., Geisler, W.S.: Multichannel texture analysis using localized spatial filters. IEEE Trans. Pattern Anal. Mach. Intell. **12**, 55–73 (1990)
11. Dalal, N., Triggs, B.: Histograms of oriented gradients for human detection. In: 2005 IEEE Computer Society Conference on Computer Vision and Pattern Recognition (CVPR 2005), vol. 1, pp. 886–893. IEEE (2005)
12. Ojala, T., Pietikainen, M., Maenpaa, T.: Multiresolution gray-scale and rotation invariant texture classification with local binary patterns. IEEE Trans. Pattern Anal. Mach. Intell. **24**, 971–987 (2002)
13. Breiman, L.: Random forests. Mach. Learn. **45**, 5–32 (2001)
14. Bishop, C.M.: Pattern recognition. Mach. Learn. **128**, 1–58 (2006)
15. Ronneberger, O., Fischer, P., Brox, T.: U-net: convolutional networks for biomedical image segmentation. In: Navab, N., Hornegger, J., Wells, W.M., Frangi, A.F. (eds.) MICCAI 2015. LNCS, vol. 9351, pp. 234–241. Springer, Cham (2015). https://doi.org/10.1007/978-3-319-24574-4_28

Video Activity Recognition Using Sequence Kernel Based Support Vector Machines

Sony S. Allappa[1], Veena Thenkanidiyoor[1(✉)], and Dileep Aroor Dinesh[2]

[1] National Institute of Technology Goa, Farmagudi, India
sonynitgoa@gmail.com, veenat@nitgoa.ac.in
[2] Indian Institute of Technology Mandi, Mandi, Himachal Pradesh, India
addileep@iitmandi.ac.in

Abstract. This paper addresses issues in performing video activity recognition using support vector machines (SVMs). The videos comprise of sequence of sub-activities where a sub-activity correspond to a segment of video. For building activity recognizer, each segment is encoded into a feature vector. Hence a video is represented as a sequence of feature vectors. In this work, we propose to explore GMM-based encoding scheme ot encode a video segment into bag-of-visual-word vector representation. We also propose to use Fisher score vector as an encoded representation for a video segment. For building SVM-based activity recognizer, it is necessary to use suitable kernel that match sequences of feature vectors. Such kernels are called sequence kernels. In this work, we propose different sequence kernels like modified time flexible kernel, segment level pyramid match kernel, segment level probability sequence kernel and segment level Fisher kernel for matching videos when segments are represented using an encoded feature vector representation. The effectiveness of the proposed sequence kernels in the SVM- based activity recognition are studied using benchmark datasets.

Keywords: Video activity recognition
Gaussian mixture Model based encoding · Fisher score vector
Support evctor machine · Time flexible kernel
Modified time flexible kernel · Segment level pyramid match kernel
Segment level probability sequence kernel · Segment level Fisher kernel

1 Introduction

Video activity recognition is one of the most interesting tasks, due to its benefits in areas such as intelligent video surveillance, automatic cinematography, elderly behavioral management, human-computer interaction, etc. Activity recognition involves assigning an activity label to a video which human beings are good at doing. However, video activity recognition is a challenging task for a computer. This is because, a video activity comprises a sequence of sub-activities. The order

M. De Marsico et al. (Eds.): ICPRAM 2018, LNCS 11351, pp. 164–185, 2019.
https://doi.org/10.1007/978-3-030-05499-1_9

in which the sub-activities appear characterizes an activity class. For example, the activities "Getting Out of Car" and "Getting Into Car" have common set of sub-activities like 'person opens car door', 'goes out of the car', 'closes door' and 'walks'. If the sequence of sub-activities are in the order "person opens car door, goes out of the car, closes door and walks", then the activity is "Getting Out of Car". If the sub-activities are in the order "person walks, opens car door, gets into the car and closes door", the activity indicates "Getting Into Car". So, the temporal ordering of sub-activities is important for discrimination between the video activities. An automatic approach to video activity recognition should consider the temporal ordering of sub-activities to discriminate one video activity from another.

A video, which is a sequence of frames can be viewed as a three dimensional matrix that carries rich spatio-temporal information. An activity recognizer should use this rich spatio-temporal information [16–18]. This is possible by extracting suitable features that capture spatio-temporal information. This leads to representing a video as a sequence of feature vectors, $\mathbf{X} = (\mathbf{x}_1, \mathbf{x}_2, .., \mathbf{x}_t, .., \mathbf{x}_T)$ where, $\mathbf{x}_t \in R^d$ and T is the length of the sequence. The length of the video sequences vary from one sequence to other because videos are of different lengths. Hence, video activity recognition involves classification of varying length sequences of feature vectors. The process of video activity recognition is illustrated in Fig. 1. It is seen from Fig. 1 that, an activity recognizer involves first extracting spatio-temporal features from a video and then using a suitable classification model to recognize the activity.

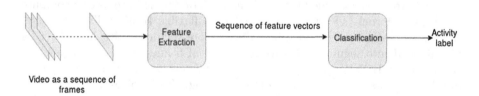

Fig. 1. Illustration of video activity recognition process.

It is necessary for a video activity recognizer to discriminate among activity classes that exhibit strong inter-class similarity. Conventionally hidden Markov model (HMM) based classifiers are used for classification of varying length sequences of feature vectors [5]. Since HMMs are built using non-discriminative learning based approaches, they may not be effective for activity recognition. Since activity classes exhibit strong inter-class similarity, a discriminative learning based approach such as a support vector machine (SVM) based classifier is expected to be helpful in building an activity recognizer. In this work, we propose to build a SVM-based classifier for activity recognition that involves classification of varying length sequences corresponding to videos.

Classification of varying length sequences of feature vectors using a SVM-based classifier requires the design of a suitable kernel as a measure of similarity between a pair of sequences. Kernels designed for varying length sequences of feature vectors are called sequence kernels [33]. In this work, we propose the design of sequence kernels for video activity recognition. While designing sequence kernels for video activity recognition, it is necessary to consider the temporal ordering of sub-activities. A sub-activity in a video corresponds to a small portion of video called a segment of the video. This segment of a video brings the contextual information that is necessary for activity recognition. So, a sequence of sub-activities correspond to sequence of segments. It is now necessary that in the design of a sequence kernel the context segments of videos should be considered. However, segmenting a video into sub-activities is not a trivial task. Instead, in this work we propose to consider the contextual segments in different ways for designing the sequence kernels. These contextual segments are approximately corresponding to sub-activities. In this work, we propose to encode a context segment of a video in two ways and then this encoded representation is used in the kernel computation. In the first approach to encoding, a segment of a video is encoded into a bag-of-visual-words (BOVW) vector. The visual words represent local semantic concepts and are obtained by clustering the feature vectors of all the video sequences of all the activity classes. The BOVW vector corresponds to histogram of visual words that occur in that segment. In the second approach, a segment of video is encoded into a Fisher score vector obtained using Gaussian mixture model (GMM) based likelihood scores. This involves first building a GMM using the feature vectors of all the video sequences of all the activity classes. A Fisher score vector is obtained using the first order derivatives of the log-likelihood with respect to the parameters of a GMM.

In this work, we extended the previous work of [36] and propose to compute time flexible kernel (TFK) and modified time flexible kernel (MTFK) where, each context segment is encoded into Fisher score vector. To design TFK [1], a video divided into segments of a fixed number of frames and every segment is encoded into a BOVW vector. As a result, a video is represented as a sequence of BOVW vectors. Matching a pair of videos using TFK involves matching every BOVW vector from a sequence with every BOVW vector from other sequence. In TFK, a pair of BOVW vectors is matched using a linear kernel (LK). In the design of MTFK [36], better approaches to match a pair of BOVW vectors are considered. The BOVW representation corresponds to frequency of occurrence of visual words. It is shown in [6] that frequency based kernels are suitable for matching a pair of frequency based vectors. In the design of MTFK, the frequency based kernels are used for matching a pair of BOVW vectors [36]. In this work, we propose to encode every segment into Fisher score vector. As a result, a video is represented as a sequence of Fisher score vectors. We then propose to compute TFK using Improved Fisher kernel (IFK) to match a pair of Fisher score vectors.

It is possible for a sequence of sub-activities at finer level to correspond to higher level sub-activities. For example, in the game of cricket, 'bowling event' may comprise of 'running' and 'throwing a ball'. Hence matching a pair of video sequences at different abstract levels of sub-activities may be helpful. We propose to explore segment level pyramid match kernel (SLPMK) [21] and segment level probabilistic sequence kernel (SLPSK) [32], used in the context of speech for activity recognition where matching between a pair of videos is done at different levels of segments. The design of SLPMK is inspired by the spatial pyramid match kernel [21]. Here, a video sequence is decomposed into pyramid of increasingly finer segments. Every segment is encoded into a BOVW vector. The SLPMK between a pair of video sequences is computed by matching the corresponding segments at each level in the pyramid. In the design of SLPSK, a video is divided into a fixed number of segments. Every segment is mapped onto a high dimensional probabilistic score space. The proposed SLPSK is computed as a combination of probabilistic sequence kernel (PSK) computed between a pair of segments which corresponds to inner product between the probabilistic score space representation.

Inspired by the concept of SLPMK, we propose segment level Fisher kernel (SLFK) for video activity recognition. In the design of SLFK, every segment is encoded as a Fisher score vector and the SLFK is computed as a combination of IFK computed between the segments of the videos. The effectiveness of the proposed kernels is studied using benchmark datasets.

The main contributions of this paper are as follows.

* GMM-based encoding scheme to encode a segment of a video into a BOVW vector representation. The GMM-based approach uses soft assignment to clusters which is found to be effective when compared to codebook based encoding which uses K-means clustering.
* Fisher score vector, as an encoded representation for a context segment of video.
* TFK and MTFK where each segment is encoded into BOVW vector representation and Fisher score vector representation.
* SLPMK, SLPSK and SLFK for matching a pair of videos at different levels of segments (or sub-activities).
* Demonstration of the effectiveness of proposed sequence kernel based SVM, with state-of-the-art results, for video activity recognition.

This paper is organized as follows. A brief overview of approaches to activity recognition is presented in Sect. 2. In Sect. 3, we present video representation techniques. Dynamic kernels used in this work are presented in Sect. 4. The experimental studies are presented in Sect. 4. In Sect. 5, we present the conclusions.

2   Video Representation

For building an effective activity recognizer, it is important to represent a video in a suitable way. This requires first to extract spatio-temporal features from a video. The sequential kernels proposed in this work need to consider matching a pair of videos using segments of videos to consider temporal ordering of sub-activities. Hence, it is useful to encode a video segment into a suitable representation. In this section, we first present an approach to feature extraction and then we present approaches to encode a video.

Video has a rich spatio-temporal information. The feature descriptors should be such that, they preserve the spatio-temporal information. Conventional approaches to video representation were extended from image representation that involves extracting features such as edges, colors, corners, etc. from every frame of video [16,18]. In this method, every frame is represented by a feature vector so that a video is represented as a sequence of feature vectors to capture temporal information. In this work, we propose to consider improved dense trajectories (IDT) based features. The IDT is a state-of-the-art descriptor, used to retain both spatial and temporal information effectively. An illustration for IDT-based feature extraction is given in Fig. 2. As shown in Fig. 2, the process densely samples feature points in each frame and tracks them in the video based on optical flow. Instead of mapping an individual frame, or a group of frames sequentially, this approach uses the sliding window method to keep the temporal information intact. We choose a window size of B frames, and sliding length of F frames. Multiple descriptors are computed along the trajectories of feature points to capture shape, appearance and motion information. The descriptors such as Histogram of Oriented Gradient (HOG), Histogram of Optical Flow (HOF) [10] and Motion Boundary Histograms (MBH) are extracted from each trajectories. The IDT descriptor is found to be effective for video analysis tasks [1]. Capturing spatio-temporal information in videos leads to representing a video as a sequence of feature vectors $\mathbf{X} = (\mathbf{x}_1, \mathbf{x}_2, ..., \mathbf{x}_t, ..., \mathbf{x}_T)$ where $\mathbf{x}_t \in R^d$ and T is the length of the sequence. Since videos are of different lengths, the lengths of the corresponding sequences of feature vectors also vary from one video to another.

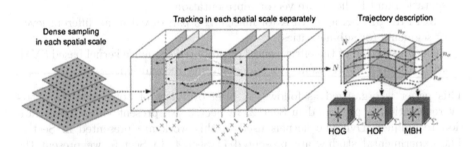

Fig. 2. An illustration for the process of improved dense trajectories (IDT) based feature extraction [10].

An activity is a sequence of sub-activities. A sub-activity corresponds to a small portion of a video called segment. It would be helpful to represent a video in terms of the segments. In this work, we propose to encode a segment of a video into a suitable representation and then design the sequence kernel using them. We propose to consider two methods for video encoding. The first method involves encoding a video segment into bag-of-visual-words (BOVW) representation. The second method encodes a segment of the video into a Fisher score vector.

BOVW Encoding: The BOVW representation of a video, $\mathbf{X}$ corresponds to a vector of frequencies of occurrence of visual words denoted by $\mathbf{z} = [z_1, z_2, ...z_k, ..., z_K]^T$. Here, z_k corresponds to the frequency of occurrence of the kth visual word in the video and K is the total number of visual words. A visual word represents a specific semantic pattern shared by a group of low-level descriptors. To obtain BOVW representation, every feature vector is assigned to the closest visual word. Visual words are obtained by clustering all the feature vectors $\mathbf{x} \in R^d$ from all the videos into K clusters. Conventionally K-means clustering approach is used to cluster the feature vectors of all the video sequences. The visual words correspond to the cluster centers and are also called as codewords. An encoding approach using a set of codewords is known as codebook based encoding (CBE). K-means clustering is a hard clustering approach. Soft cluster assignment is expected to be better in the encoding process. In this work, we propose to consider Gaussian mixture model (GMM) based approach for soft clustering. This involves building a GMM and every component of GMM is considered as a representation for a visual word. The video encoding using a GMM is called as GMM-based encoding (GMME). The CBE and GMME for video encoding that are presented below.

Codebook Based Encoding: In the codebook based encoding (CBE) scheme a feature vector is assigned with the index of the closest visual word using Euclidean distance as follows:

$$m = \underset{k}{\mathrm{argmin}} ||\mathbf{x} - \boldsymbol{\mu}_k|| \tag{1}$$

Here, $\boldsymbol{\mu}_k$ denotes the kth visual word that corresponds to the center of kth cluster. In this method only the centers of the clusters are used. Information such as spread of the cluster and the strength of the cluster are not considered.

Gaussian Mixture Model Based Encoding: In this method, a feature vector can belong to more than one cluster with non-zero probability. In this method, components of the GMM correspond to the visual words. The belongingness of a feature vector $\mathbf{x}$ to a cluster k in the GMM-based encoding (GMME), is given by the responsibility term,

$$\gamma_k(\mathbf{x}) = \frac{w_k \mathcal{N}(\mathbf{x}|\boldsymbol{\mu}_k, \boldsymbol{C}_k)}{\sum_{i=1}^{K} w_i \mathcal{N}(\mathbf{x}|\boldsymbol{\mu}_i, \boldsymbol{C}_i)} \tag{2}$$

where w_k is the mixture coefficient of the component k, and $\mathcal{N}(\mathbf{x}|\boldsymbol{\mu}_k, \boldsymbol{C}_k)$ is the Gaussian density for the component k with mean vector $\boldsymbol{\mu}_k$ and covariance matrix $\boldsymbol{C}_k$. For a video sequence with T feature vectors, the z_k is computed as,

$$z_k = \sum_{t=1}^{T} \gamma_k(\mathbf{x}_t) \tag{3}$$

Our studies and experimental results show that the performance obtained for activity recognizer that uses GMME for representing a video is better compared to that uses CBE method. In the encoding process presented so far, entire video is encoded into a BOVW representation that corresponds to a histogram of visual words. An important limitation of encoding entire video, $\mathbf{X}$ into $\mathbf{z}$ is that the temporal information among the frames of the video is lost. As video activity is a very complex phenomena that involves various sub-activities, the temporal information corresponding to the sub-activities in an activity is essential for video activity recognition. The sub-activities also correspond to segments of a video and the ordering of these segments is important. The segments bring contextual information. To build an activity recognizer, it is helpful to match a pair of videos at segment level. For this each segment can be encoded into a BOVW vector. This requires to split a video into a sequence of segments and encode every segment into a BOVW vector. This results in a sequence of BOVW vectors representation for a video, $\mathbf{Y} = (\mathbf{y}_1, \mathbf{y}_2, ..., \mathbf{y}_n, ..., \mathbf{y}_N)$ where $\mathbf{y_n} \in R^K$. Here, N corresponds to the number of video segments considered.

Fisher Encoding: In this work, we also propose to encode a video segment as a Fisher score vector.

Let $\mathbf{X} = (\mathbf{x}_1, \mathbf{x}_2, ..\mathbf{x}_t, ...\mathbf{x}_T)$, where $\mathbf{x}_t \in R^d$ denote a segment of a video. Let $\gamma_{kt}, k = 1, \ldots, K, t = 1, \ldots, T$ denote the soft assignments of the T feature vectors of a video segment to K Gaussian components. For each $k = 1, \ldots, K$, define the vectors

$$\mathbf{u}_k = \frac{1}{N\sqrt{w_k}} \sum_{t=1}^{T} \gamma_{kt} \mathbf{C}_k^{-1/2}(\mathbf{x}_t - \boldsymbol{\mu}_k) \tag{4}$$

$$\mathbf{v}_k = \frac{1}{N\sqrt{2w_k}} \sum_{t=1}^{T} \gamma_{kt}[(\mathbf{x}_t - \boldsymbol{\mu}_t)\mathbf{C}_k^{-1}(\mathbf{x}_t - \boldsymbol{\mu}_k) - 1] \tag{5}$$

The Fisher encoding of the sequence of feature vector is then given by the concatenation of $\mathbf{u}_k$ and $\mathbf{v}_k$ for all K components.

$$\mathbf{f} = [\mathbf{u}_1^T, \mathbf{v}_1^T, \ldots, \mathbf{u}_K^T, \mathbf{v}_K^T]^T \tag{6}$$

To take into consideration the temporal ordering of video sub-activities, we propose to encode an activity video into a sequence of Fisher vectors as $\mathbf{F} = (\mathbf{f}_1, \mathbf{f}_2, ...\mathbf{f}_n, ..., \mathbf{f}_N)$.

Video activity recognition is a challenging task due to the high level of inter-class similarity exhibited by the activity classes. It is very essential for an activity recognizer to discriminate among the activity classes. Support vector machines (SVMs) are shown to be effective in building discriminative classifiers. For building an effective activity recognizer using SVMs, it is necessary to use suitable kernels. We present the kernels proposed in this work in the next section.

3   Sequence Kernels for Video Activity Recognition

Activity recognition in videos involves considering the sequence of feature vectors $\mathbf{X} = (\mathbf{x}_1, \mathbf{x}_2, ...\mathbf{x}_t, ...\mathbf{x}_T)$ representation of videos. Here, $\mathbf{x}_t \in \mathbf{R}^d$ and $\mathbf{T}$ is the length of the sequence. Videos being different in length, the length of the sequence varies from one video to another. To build an SVM-based activity recognizer, it is necessary to consider suitable kernels that consider varying length sequences of feature vectors. Kernels that consider varying length sequences of feature vectors are known as sequence kernels [33]. An activity video is a sequence of sub-activities. The temporal ordering of sub-activities is important for activity recognition. A sub-activity corresponds to a small portion of a video called segment. The segments bring contextual information for building an activity recognizer. We propose to design sequence kernels that consider these contextual segments and match a pair of video at the segment level. In this section, we propose the design of five sequence kernels namely, time flexible kernel (TFK), modified time flexible kernel (MTFK), segment level pyramid match kernel (SLPMK), segment level probabilistic sequence kernel (SLPSK) and segment level Fisher kernel (SLFK).

3.1   Time Flexible Kernel

To design a time flexible kernel (TFK), a sequence of feature vectors, $\mathbf{X} = (\mathbf{x}_1, \mathbf{x}_2, ..., \mathbf{x}_t, ..., \mathbf{x}_T)$ is first divided into sequence of segments such that, $\mathbf{X} = (\mathbf{X}^1, \mathbf{X}^2, ..., \mathbf{X}^n, ..., \mathbf{X}^N)$, where, $\mathbf{X}^n$ corresponds to n^{th} segment and N is the number of segments. A segment is obtained by considering a sliding window of $'B'$ frames. Every segment $\mathbf{X}^n$ is encoded into a bag-of-visual-word (BOVW) vector $\mathbf{y}_n \in R^K$ where, $\mathbf{y}_n = [y_{n1}, y_{n2}, .., y_{nk}, .., y_{nK}]^T$. This results in encoding a sequence of feature vectors, $\mathbf{X} = (\mathbf{x}_1, \mathbf{x}_2, ..., \mathbf{x}_t, ..., \mathbf{x}_T)$ into a sequence of BOVW vectors, $\mathbf{Y} = (\mathbf{y}_1, \mathbf{y}_2, .., \mathbf{y}_n, .., \mathbf{y}_N)$.

Let $\mathbf{Y}_i = (\mathbf{y}_{i1}, \mathbf{y}_{i2}, ..., \mathbf{y}_{in}, ..., \mathbf{y}_{iN})$ and $\mathbf{Y}_j = (\mathbf{y}_{j1}, \mathbf{y}_{j2}, ..., \mathbf{y}_{jm}, ..., \mathbf{y}_{jM})$ correspond to the sequence of BOVW vectors representation of ith and j^{th} videos respectively. Here, N and M represent the length of the sequence of BOVW vectors. The TFK involves matching every BOVW vector from $\mathbf{Y}_i$ with every BOVW vector from $\mathbf{Y}_j$. It was observed that in an activity video, middle of the video has the core of the activity [1] and it is necessary to ensure maximum matching at the middle of the video. Hence to effectively match a pair of videos, it is necessary to ensure that their centers are alligned. This is achieved by using a weight, w_{nm} for matching between y_{in} and y_{jm}. The value of w_{nm} is large when $n = N/2$ and $m = M/2$, i.e., matching at the center of two sequences. The value of w_{nm} for $n = N/2, m = 1$ will be smaller than w_{nm} for $n = N/2, m = M/2$. The details of choosing w_{nm} can be found in [1]. Effectively the TFK is a weighted summation kernel as given below:

$$K_{\text{TFK}}(\mathbf{Y}_i, \mathbf{Y}_j) = \sum_{n=1}^{N} \sum_{m=1}^{M} w_{nm} K_{\text{LK}}(\mathbf{y}_{in}, \mathbf{y}_{jm}) \qquad (7)$$

An illustration of matching a BOVW vector $\mathbf{y}_{j9}$ from the sequence $\mathbf{Y}_j$ with all the BOVW vectors of $\mathbf{Y}_i$ is given in Fig. 3. In this illustration, length of $\mathbf{Y}_i$ is 11 and that of $\mathbf{Y}_j$ is 17. For every match between $\mathbf{y}_{j9}$ with the BOVW vectors from $\mathbf{Y}_i$, a weight $w_{9i}, i = 1, 2, ...11$ is considered. This is because $\mathbf{y}_{j9}$ and $\mathbf{y}_{i6}$ correspond to the center of activity videos where the match needs to be maximum. In (10), a linear kernel is used for matching $\mathbf{y}_{in}$ and $\mathbf{y}_{jm}$. In principle any kernels on fixed-length representation of examples can be used in place of $K_{LK}(\mathbf{y}_{in}, \mathbf{y}_{jm})$. It is also helpful if better ways of matching a pair of BOVW vectors can be explored. In the nect section we present modified time flexible kernel (MTFK).

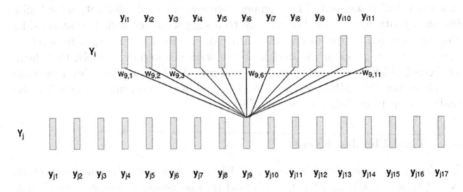

Fig. 3. An illustration of matching $\mathbf{y}_{j9}$ from $\mathbf{Y}_j$ with all the BOVW vectors from $\mathbf{Y}_i$ using suitable weights. Here, the length of $\mathbf{Y}_i$ is 11 and that of $\mathbf{Y}_j$ is 17.

3.2   Modified Time Flexible Kernel

In the design of modified time flexible kernel, we propose to explore better ways of matching a pair of BOVW vectors. The widely used non linear kernels on fixed-length representations such as the Gaussian kernel (GK) or the polynomial kernel (PK) can be used. However, the computation of these kernels require fine tuning of kernel parameters which is not easy. In this work, each video segment is represented using BOVW representation which is a histogram vector representation. For such representations, the frequency-based kernels for SVMs are found to be more effective [6]. So, we have explored two frequency based kernels namely, Histogram Intersection Kernel (HIK) [7] and Hellinger's Kernel (HK) [8] as non linear kernels to modify the TFK.

HIK-Based Modified TFK. Let $\mathbf{y}_{in} = [y_{in1}, y_{in2}, ..., y_{inK}]^T$ and $\mathbf{y}_{jm} = [y_{jm1}, y_{jm2}, ..., y_{jmK}]^T$ be the nth and m^{th} elements of the sequence of BOVW vectors $\mathbf{Y}_i$ and $\mathbf{Y}_j$ corresponding to the video sequences $\mathbf{X}_i$ and $\mathbf{X}_j$ respectively.

The number of matches in the kth bin of the histogram is given by histogram intersection function as,

$$s_k = min(y_{ink}, y_{jmk}) \tag{8}$$

HIK is computed as the total number of matches given by [7],

$$K_{\text{HIK}}(\mathbf{y}_{in}, \mathbf{y}_{jm}) = \sum_{k=1}^{K} s_k \tag{9}$$

HIK-based modified TFK is given by,

$$K_{\text{HIKMTFK}}(\mathbf{Y}_i, \mathbf{Y}_j) = \sum_{n=1}^{N} \sum_{m=1}^{M} w_{nm} K_{\text{HIK}}(\mathbf{y}_{in}, \mathbf{y}_{jm}) \tag{10}$$

HK-Based Modified TFK. In Hellinger's kernel the number of matches in the kth bin of the histogram given by,

$$s_k = \sqrt{y_{ink} y_{jmk}} \tag{11}$$

Hellinger's kernel is computed as the total number of matches across the histogram. It is given by,

$$K_{\text{HK}}(\mathbf{y}_{in}, \mathbf{y}_{jm}) = \sum_{k=1}^{K} s_k \tag{12}$$

HK-based modified TFK is given by:

$$K_{\text{HKMTFK}}(\mathbf{Y}_i, \mathbf{Y}_j) = \sum_{n=1}^{N} \sum_{m=1}^{M} w_{nm} K_{\text{HK}}(\mathbf{y}_{in}, \mathbf{y}_{jm}) \tag{13}$$

When video segment is encoded into BOVW vector, we have proposed HK-based MTFK and HIK-based MTFK. When the encoding scheme is Fisher encoding, we propose improved Fisher kernel (IFK) based modified TFK.

IFK-Based Modified TFK. Let $\mathbf{F}_i = (\mathbf{f}_{i1}, \mathbf{f}_{i2}, ...\mathbf{f}_{in}, ..., \mathbf{f}_{iN})$ and $\mathbf{F}_j = (\mathbf{f}_{j1}, \mathbf{f}_{j2}, ...\mathbf{f}_{jn}, ..., \mathbf{f}_{jN})$ be the sequence of Fisher score vector representation for i^{th} and j^{th} video. The improved Fisher kernel between $\mathbf{f}_{in}$ and $\mathbf{f}_{jm}$ is computed as,

$$K_{\text{IFK}}(\mathbf{f}_{in}, \mathbf{f}_{jm}) = \mathbf{f}_{in}^T \mathbf{f}_{jm} \tag{14}$$

IFK-based modified TFK is given by,

$$K_{\text{IFMTFK}}(\mathbf{F}_i, \mathbf{F}_j) = \sum_{n=1}^{N} \sum_{m=1}^{M} w_{nm} K_{\text{IFK}}(\mathbf{f}_{in}, \mathbf{f}_{jm}) \tag{15}$$

In the design of TFK and MTFK, a video is represented as a sequence of segments. It is possible for a sequence of sub-activities occurring one after the other to form a higher level of sub-activity. For example, in the video of cricket game, sub-activities such as 'running' and 'throwing a ball' form a higher level sub-activity, 'bowling'. Hence, it may be desirable to match a pair of videos at different levels of sub-activities. For this, we present segment level pyramid match kernel in the next section.

Combining Kernels. To take the maximum advantage of the video representation, we consider the BOVW encoding of the entire video $\mathbf{z}_i$ and the sequence of BOVW representation, $\mathbf{Y}_i$ of a video sequence, $\mathbf{X}_i$. We consider linear combination of kernels, MTFK and kernel computed on BOVW encoding of entire video.

$$K_{COMB}(\mathbf{X}_i, \mathbf{X}_j) = K_1(\mathbf{Y}_i, \mathbf{Y}_j) + K_2(\mathbf{z}_i, \mathbf{z}_j) \qquad (16)$$

Here, $K_1(\mathbf{Y}_i, \mathbf{Y}_j)$ is either HIK-based MTFK or HK-based MTFK and $K_2(\mathbf{z}_i, \mathbf{z}_j)$ is LK or HIK or HK.

The base kernel (LK or HIK or HK) is a valid positive semidefinite kernel and multiplying a valid positive semidefinite kernel by a scalar is a valid positive semidefinite kernel [15]. Also, the sum of valid positive semidefinite kernels is a valid positive semidefinite kernel [15]. Hence, both TFK and modified TFK are valid positive semidefinite kernels.

3.3   Segment Level Pyramid Match Kernel

An activity video is a sequence of sub-activities. To design segment level pyramid match kernel (SLPMK), a video is decomposed into increasingly finer segments and is represented as a pyramid of segments. To compute segment level pyramid match kernel (SLPMK) between two videos, we match the corresponding video segments at each level of the pyramid. Let $l = 0, 1, ..., L-1$ be the L levels of the pyramid [32]. At 0th level, complete video sequence is considered as a segment. At the 1st level, video sequence is divided into two equal segments. At the 2nd level, a video sequence is divided into four equal segments and so on. Hence, at any level l, a video sequence is divided into 2^l equal segments. Every segment is encoded into a BOVW vector before matching. In this work, we propose to use two approaches to encode a video segment into a BOVW vector. In the first approach codebook based encoding (CBE) is used. In the second approach, GMM-based encoding (GMMe) is used. The design of SLPMK when CBE and GMME are used is presented as follows.

Codebook Based SLPMK. Let $\mathbf{X}_i$ and $\mathbf{X}_j$ be the ith and j^{th} videos to be matched using the proposed SLPMK. At the lth level of the pyramid, let $\mathbf{y}_{lp}$ be the K-dimensional BOVW vector corresponding to p^{th} segment. Let y_{lpk} be the kth element of $\mathbf{y}_{lp}$ that corresponds to the number of feature vectors of pth segment assigned to the kth codeword. The number of matches in the kth codeword between the pth segments of $\mathbf{X}_i$ and $\mathbf{X}_j$ at l^{th} level of pyramid is given by,

$$s_{lpk} = min(y_{ilpk}, y_{jlpk}) \qquad (17)$$

Total number of matches at level l between the pth segments of $\mathbf{X}_i$ and $\mathbf{X}_j$ is given by,

$$S_{lp} = \sum_{k=1}^{K} s_{lpk} \tag{18}$$

Total number of matches between $\mathbf{X}_i$ and $\mathbf{X}_j$ at level l is obtained as,

$$\hat{S}_l = \sum_{p=1}^{2^l} S_{lp} \tag{19}$$

The number of matches found at level l also includes all the matches found at the finer level $l + 1$. Therefore, the number of new matches found at level l is given by $\hat{S}_l - \hat{S}_{l+1}$. The codebook based segment level pyramid match kernel (CBSLPMK) is computed as:

$$K_{\mathrm{CBSLPMK}}(\mathbf{X}_i, \mathbf{X}_j) = \sum_{l=0}^{L-2} \frac{1}{2^{L-(l+1)}} (\hat{S}_l - \hat{S}_{l+1}) + \hat{S}_{L-1} \tag{20}$$

In this method, K-means clustering is used to construct bag-of-codewords. Soft clustering could be used to construct a better SLPMK. In the next sub-section, we use GMM-based SLPMK. Information about the spread, the size of clusters along with the centers of clusters is considered in GMM- based soft assignment of feature vectors.

GMM-Based SLPMK. Here, a segment of a video is encoded using GMME. This involves building a GMM using all the feature vectors of all the sequences of all the activity classes. The soft assignment of a feature vector $\mathbf{x}_t$ of a segment of a video to the kth component of GMM is given by the responsibility term,

$$\gamma_k(\mathbf{x}_t) = \frac{w_k \mathcal{N}(\mathbf{x}_t | \boldsymbol{\mu}_k, \boldsymbol{C}_k)}{\sum_{j=1}^{K} w_j \mathcal{N}(\mathbf{x}_t | \boldsymbol{\mu}_j, \boldsymbol{C}_j)} \tag{21}$$

where, $\mathcal{N}(\mathbf{x}_t | \boldsymbol{\mu}_k, \boldsymbol{C}_k)$ is the kth Gaussian component with mean vector $\boldsymbol{\mu}_k$ and covariance matrix $\boldsymbol{C}_k$. Here, w_k denotes the mixture weight. For the pth video segment at l^{th} level of pyramid, the effective number of feature vectors of a sequence $\mathbf{X}$ assigned to the component k given by,

$$y_{lpk} = \sum_{t=1}^{T} \gamma_k(\mathbf{x}_t) \tag{22}$$

where, T is the number of feature vectors in the pth segment of $\mathbf{X}$. For a pair of examples represented as sequence of feature vectors, $\mathbf{X}_i$ and $\mathbf{X}_j$, number of matches in the kth codeword between the pth segments of $\mathbf{X}_i$ and $\mathbf{X}_j$ at l^{th} level of pyramid (s_{lpk}), total number of matches at level l between the pth segments (s_{lp}) and total number of matches between $\mathbf{X}_i$ and $\mathbf{X}_j$ at level l ($\hat{S}_l$) are computed as in (17), (18) and (19) respectively. GMMSLPMK between a pair of videos $\mathbf{X}_i$ and $\mathbf{X}_j$, K_{GMMSLPMK} is then computed as in (20). In the next section, we present segment level probabilistic sequence kernel.

3.4   Segment Level Probabilistic Sequence Kernel

Segment level probabilistic sequence kernel (SLPSK) [32] divides a video sequence into a fixed number of segments and then maps each segment onto a probabilistic feature vector. SLPSK between a pair of videos is obtained by matching the corresponding segments. Let $\mathbf{X}_i = (\mathbf{x}_{i1}, \mathbf{x}_{i2}, ...\mathbf{x}_{it}, ...\mathbf{x}_{iT_i})$ and $\mathbf{X}_j = (\mathbf{x}_{j1}, \mathbf{x}_{j2}, ...\mathbf{x}_{jt}, ...\mathbf{x}_{jT_j})$ be the sequence of feature vectors representation corresponding to i^{th} and j^{th} video. Let $\mathbf{X}_i$ and $\mathbf{X}_j$ be divided into N segments, such that $\mathbf{X}_i = (\mathbf{X}_i^1, \mathbf{X}_i^2, ...\mathbf{X}_i^n, ...\mathbf{X}_i^N)$ and $\mathbf{X}_j = (\mathbf{X}_j^1, \mathbf{X}_j^2, ...\mathbf{X}_j^n, ...\mathbf{X}_j^N)$ be the sequence of segments corresponding to $\mathbf{X}_i$ and $\mathbf{X}_j$ respectively. Here, let $\mathbf{X}_i^n = (\mathbf{x}_{i1}^n, \mathbf{x}_{i2}^n, ...\mathbf{x}_{it}^n, ...\mathbf{x}_{iT_{in}}^n)$ be the sequence of feature vectors corresponding to the nth segment in the ith video. In the design of SLPSK, probabilistic sequence kernel (PSK) is computed between corresponding segments of the two videos.

The PSK uses universal background model (UBM) with K components and the class-specific GMMs obtained by adapting to the UBM. The UBM, also called as class independent GMM (CIGMM), is a large GMM built using the training data of all the classes. A feature vector $\mathbf{x}_{it}^n$ corresponding to the tth feature vector in the nth segment of ith video, is represented in a higher dimensional feature space as a vector of responsibility terms of the $2K$ components (K from class-specific adapted GMM and other K from UBM). $\Psi(\mathbf{x}_{it}^n) = [\gamma_1(\mathbf{x}_{it}^n), \gamma_2(\mathbf{x}_{it}^n), ..., \gamma_{2K}(\mathbf{x}_{it}^n)]^T$. Since the element $\gamma_k(\mathbf{x}_{it}^n)$ indicates the probabilistic alignment of $\mathbf{x}_{it}^n$ to the kth component, $\Psi(\mathbf{x}_{it}^n)$ is called the probabilistic alignment vector which includes the information common to all the classes. A sequence of feature vectors $\mathbf{X}_i^n$ corresponding to the nth segment of the ith video is represented as a fixed dimensional vector $\Phi_{PSK}^n(\mathbf{X}_i^n)$, and is given by,

$$\Phi_{PSK}^n(\mathbf{X}_i^n) = \frac{1}{T_{in}} \sum_{i=1}^{T_{in}} \Psi(\mathbf{x}_{it}^n) \tag{23}$$

Then, the PSK between two segments $\mathbf{X}_i^n$ and $\mathbf{X}_j^n$ is computed as,

$$K_{PSK}^n(\mathbf{X}_i^n, \mathbf{X}_j^n) = \Phi_{PSK}^n(\mathbf{X}_i^n)^T S_n^{-1} \Phi_{PSK}^n(\mathbf{X}_j^n) \tag{24}$$

where, S_n is the correlation matrix given by,

$$S_n = \frac{1}{M_n} R_n^T R_n \tag{25}$$

where, R_n is the matrix whose rows are the probabilistic alignment vectors for feature vectors of the nth segment and M_n is the total number of feature vectors in the nth segment.

The SLPSK between $\mathbf{X}_i$ and $\mathbf{X}_j$ is then computed as combination of the segment-specific PSKs as follows,

$$K_{SLPSK}(\mathbf{X}_i, \mathbf{X}_j) = \sum_{n=1}^{N} K_{PSK}^n(\mathbf{X}_i^n, \mathbf{X}_j^n) \tag{26}$$

Since, PSK is a valid semidefinite kernel, the segment specific PSK is also a valid positive semidefininte kernel. Hence, SLPSK is also a valid positive semidefinite kernel because the sum of valid positive semidefinite kernel is a valid positive semidefinite kernel. In the next section, we present segment level Fisher kernel.

3.5   Segment Level Fisher Kernel

To design segment level Fisher kernel (SLFK), a video represented as a sequence of feature vectors is divided into a fixed number of segments. Computation of SLFK involves computing Fisher kernel (FK) between the corresponding segments of the two video sequences and then combining the segment level match to get SLFK between the two videos. Computation of FK between a pair of segments involves first encoding the segments into respective Fisher score vectors [23]. Let $\mathbf{X}_i^n = (\mathbf{x}_{i1}^n, \mathbf{x}_{i2}^n, ...\mathbf{x}_{it}^n, ...\mathbf{x}_{iT_{in}}^n)$ and $\mathbf{X}_j^n = (\mathbf{x}_{j1}^n, \mathbf{x}_{j2}^n, ...\mathbf{x}_{jt}^n, ...\mathbf{x}_{jT_{jm}}^n)$ be the nth segment of the ith and j^{th} videos. Let $\mathbf{X}_i^n$ and $\mathbf{X}_j^n$ be encoded into the Fisher score vectors $\mathbf{f}_i^n$ and $\mathbf{f}_j^n$ respectively.

The improved Fisher kernel (IFK) between two segments of the videos $\mathbf{X}_i^n$ and $\mathbf{X}_j^n$ is computed as,

$$K_{\mathrm{FK}}^n(\mathbf{X}_i^n, \mathbf{X}_j^n) = (\mathbf{f}_i^n)^T(\mathbf{f}_j^n)^T \tag{27}$$

The SLFK for $\mathbf{X}_i$ and $\mathbf{X}_j$ is then computed as combination of the segment-specific FKs as follows,

$$K_{\mathrm{SLFK}}(\mathbf{X}_i, \mathbf{X}_j) = \sum_{n=1}^{N} K_{\mathrm{FK}}^n(\mathbf{X}_i^n, \mathbf{X}_j^n) \tag{28}$$

Since, FK is a valid semidefinite kernel, the segment specific FK is also a valid positive semidefininte kernel. Hence, SLFK is also a valid positive semidef- inite kernel because the sum of valid positive semidefinite kernel is a valid positive semidefinite kernel. In the next section, we present combination of kernels.

4   Experimental Studies

In this section, we present the results of experimental studies carried out to verify the effectiveness of sequence kernels for video activity recognition. We first present the datasets used and video representation considered. Then, we present the various studies conducted.

4.1   Datasets

UCF Sports Dataset. This dataset comprises a collection of 150 sports videos of 10 activity classes. On an average, each class contains about 15 videos with an average length of each video being approximately 6.4 s. We follow leave-one-out cross validation strategy as used in [35].

UCF50 Dataset. UCF50 is an action recognition data set with 50 action categories, consisting of realistic videos taken from youtube. This data set is an extension of YouTube Action data set. This dataset contains 6681 video clips of 50 different activities. We follow leave-one-group-out cross-validation strategy used in [34].

Hollywood 2 Dataset. This dataset contains 1707 video clips belonging to 12 classes of human actions. It has 823 videos in the training dataset and 884 videos in the test dataset as used in [31].

4.2   Video Representation

Each video is represented using improved dense trajectories (IDT) descriptor [10]. IDT descriptor densely samples feature points in each frame and tracks them in the video based on optical flow. To incorporate the temporal information, IDT descriptor is extracted using a sliding window of 30 frames with an overlap of 15 frames. For a particular sliding window, multiple IDT descriptors, each of 426 dimensions are extracted. The 426 features of an IDT descriptor comprise multiple descriptors such as histogram of oriented gradient (HOG), histogram of optical flow (HOF), and motion boundary histograms (MBH). The number of descriptors per window depends on the number of feature points tracked in that window. A video clip is represented as a sequence of IDT descriptors. We propose to encode sequence of feature vectors of a video in two ways. In the first approach, entire video is encoded into a bag-of-visual-words (BOVW) representation. In the second approach, a video is encoded into a sequence of BOVW representation. In this work, we propose to consider the GMM-based soft clustering approach for video encoding. We also compare the GMM-based encoding (GMME) approach with the codebook-based encoding (CBE) approach proposed in [1]. In this work, we study the effectiveness of different sequence kernels for activity recognition using SVM-based classifiers.

4.3   Studies on Activity Recognition in Video Using BOVW Representation Corresponding to Entire Videos

In this section, we study the SVM-based activity recognition by encoding an entire video clip into a single BOVW vector representation. We consider codebook-based encoding (CBE) [1] and GMM-based encoding (GMME) methods for generating BOVW representation. Different values for the codebook size in CBE and the number of clusters in GMM, is explored and an optimal value of 256 is chosen. For SVM-based classifier, we need a suitable kernel. We propose to consider linear kernel (LK), histogram intersection kernel (HIK) and Hellinger's kernel (HK). The accuracy of activity recognition in videos using SVM-based classifier for the three datasets is given in Table 1. It is seen from Table 1 that video activity recognition using SVM-based classifier that uses the frequency based kernels, HIK and HK, is better than that using LK. This shows the suitability of frequency based kernels when the videos are encoded into BOVW representation. It is also seen that the performance of SVM-based classifier using GMME is better than that using CBE used in [1]. This shows the effectiveness of GMM-based soft clustering approach for video encoding. In the next section, we study SVM-based approach to video activity recognition when a video clip is represented as a sequence of BOVW representation.

Table 1. Accuracy in (%) of SVM-based classifier for activity recognition in videos using linear kernel (LK), Hellinger's kernel (HK) and histogram intersection kernel (HIK) on the BOVW encoding of the entire video. Here CBE corresponds to code book based encoding proposed in [1] and GMME corresponds to GMM-based video encoding proposed in this work.

| | Datasets | | | | | |
| | UCF Sports | | UCF50 | | Hollywood | |
Kernel/Encoding	CBE	GMME	CBE	GMME	CBE	GMME
LK	80.67	90.40	80.40	90.40	80.40	90.40
HK	83.33	92.93	83.33	92.27	83.00	93.57
HIK	84.67	92.53	82.27	92.53	83.00	93.13

4.4   Studies on Activity Recognition in Videos Using Sequence of Feature Vectors Representation for Videos

In this section, we study the activity recognition in videos using sequence of feature vectors representation of videos. To build an SVM-based activity recognizer, we propose to consider time flexible kernel (TFK), modified time flexible kernel (MTFK), segment level pyramid match kernel (SLPMK), segment level probabilistic sequence kernel (SLPSK) and segment level Fisher kernel (SLFK). For computing TFK, a sequence of feature vectors is encoded into a sequence of BOVW vectors where a segment of 30 frames is considered. We consider both codebook based encoding (CBE) and GMM-based encoding (GMME) approaches. Different values for the codebook size in CBE and the number of clusters in GMM, is explored and an optimal value of 256 is chosen. MTFK between a pair of videos is computed using the same sequence of BOVW vectors representation used in TFK. Here, we consider the frequency based kernels HK and HIK for MTFK. The accuracy of SVM-based activity recognizer for the three datasets using the proposed kernels is given in Table 2. It is seen from Table 2 that, MTFK-based SVMs give better performance than TFK-based SVMs. This shows the effectiveness of the kernels proposed in this work. From Tables 1 and 2, it is also seen that the TFK-based SVMs give better performance than LK-based SVMs that use BOVW vector representation of entire video. This shows the importance of using the temporal information of video for the activity recognition. It is also seen that the performance of MTFK-based SVMs is better than the SVM-based classifiers using the frequency based kernels, HK and HIK, on BOVW vectors encoding corresponding to the entire video.

It is seen from Table 2 that the performance of MTFK-based activity recognizer is comparable to that obtained using the activity recognizers that use SLPMK, SLPSK and SLFKs for UCF Sports and UCF50 datasets. In case of Hollywood dataset, SLFK based SVM classifier is found to perform better than the other classifiers. For SLPMK we have considered levels $L = 1, 2, 3$ and at each level a segment of video is encoded using CBE and GMME. Different values for the codebook size in CBE and the number of clusters in GMM, is explored and an optimal value of 256 is chosen. For SLPSK and SLFK, we divide the video into segments in 3 ways. The first approach involves dividing a video into two segments. In the second approach, a video is divided into 4 segments and in the third approach there are 8 segments considered.

Table 2. Accuracy in (%) of SVM-based classifier for activity recognition in videos using TFK, MTFK, SLPMK, SLPSK and SLFK computed on sequence of BOVW vectors representation of videos. Here, CBE corresponds to code book based encoding proposed in [1] and GMME corresponds to GMM-based video encoding proposed in this work. HK-MTFK corresponds to HK-based modified TFK and HIK-MTFK denotes the HIK-based modified TFK. SLPMK with three levels $L = 1, 2, 3$ is considered. Here SLPSK-1 denotes SLPSK computed by diving a video into two segments. SLPSK-2 and SLPSK-3 correspond to SLPSK computed by diving a video into 4 and 8 segments respectively. SLFK-1, SLFK-2 and SLFK-3 correspond to SLFK computed by diving a video into 2, 4 and 8 segments respectively.

	Datasets					
	UCF Sports		UCF50		Hollywood	
Kernel/Encoding	CBE	GMME	CBE	GMME	CBE	GMME
TFK	82.00	91.27	81.67	91.27	82.67	91.67
HK–MTFK	86.67	95.73	86.67	95.27	86.13	92.26
HIK–MTFK	86.00	95.60	86.00	95.00	86.00	92.00
SLPMK ($L = 1$)	84.50	95.60	85.07	94.06	85.00	94.00
SLPMK ($L = 2$)	85.00	96.00	85.60	94.50	85.60	94.50
SLPMK ($L = 3$)	84.50	96.00	85.67	95.00	83.00	94.07
FK–MTFK	95.67		95.00		95.07	
SLPSK-1	92.67		93.00		92.67	
SLPSK-2	93.67		94.00		94.00	
SLPSK-3	93.27		93.70		93.00	
SLFK-1	95.00		94.60		94.00	
SLFK-2	95.27		94.67		95.00	
SLFK-3	95.07		94.67		95.00	

4.5 Studies on Activity Recognition in Video Using Combination of Kernels

In this section, we combine a kernel computed on BOVW representation of entire videos and a kernel computed on sequence of BOVW vectors' representation of video. We consider simple additive combination so that a combined kernel $COMB(K_1 + K_2)$ corresponds to addition of K_1 and K_2. Here K_1 corresponds to kernel computed using sequence of BOVW vectors representation of videos, and K_2 corresponds to the kernel computed on the BOVW representation of entire video. The performance of SVM-based classifier using combined kernels for video activity recognition is given in Table 3. It is seen from Table 3 that the accuracy of SVM-based classifier using the combination kernel involving TFK is better than that for the SVM-based classifier using only TFK. This shows the effectiveness of the combination of kernels. It is also seen that the performance for SVM-based classifier using combination of kernels involving HK and HIK computed on entire video is better than that obtained with combination of kernel involving LK computed on entire video. It is also seen that the performance of SVM-based classifiers using combination kernel involving MTFKs is better than that using TFKs. This shows the effectiveness of the proposed MTFK in activity recognition in videos. In Figs. 4, 5 and 6 we compare the performance of SVM-based activity recognition using different kernels for the UCF sports dataset, UCF 50 dataset and Hollywood datasets respectively. It is seen from Figs. 4, 5 and 6 that, the SVM-based activity recognizers using the combination of kernels performs better than classifiers that use other kernels.

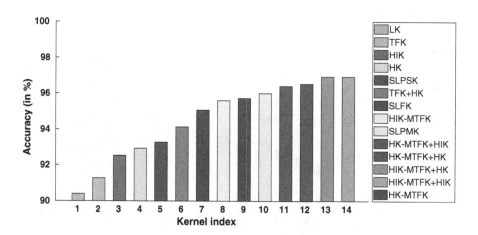

Fig. 4. Comparison of performance of SVM-based classifier using different kernels for the video activity recognition in UCF Sports dataset.

Table 3. Comparison of performance of activity recognition in video using SVM-based classifier that uses combination of kernels. Here, $COMB(K_1 + K_2)$ indicate additive combination of kernels K_1 and K_2 respectively. K_1 is a kernel computed on the sequence of BOVW vectors representation of videos and K_2 is a kernel computed on the BOVW representation of the entire video. Here, CBE corresponds to code book based encoding proposed in [1] and GMME corresponds to GMM-based video encoding proposed in this work.

Kernel/Encoding	Datasets					
	UCF Sports		UCF50		Hollywood	
	CBE	GMME	CBE	GMME	CBE	GMME
COMB(TFK+LK)	82.67	93.87	82	93.87	82.67	93.87
COMB(HK-MTFK+LK)	84.67	94.13	84.67	93.87	83.13	95
COMB(HIK-MTFK+LK)	82.67	94.13	83.67	93.87	83	95.13
COMB(TFK+HK)	86	94.13	85.13	94.13	85.13	94.26
COMB(HK-MTFK+HK)	86.67	96.53	86	96.13	86.26	96.67
COMB(HIK-MTFK+HK)	87.33	96.93	86.93	96.27	86	96.13
COMB(TFK+HIK)	84	94.67	84	94.67	84.26	94.13
COMB(HK-MTFK+HIK)	86.67	96.4	86.67	96.4	86.67	96
COMB(HIK-MTFK+HIK)	86	96.93	86.27	96.27	86	96.27

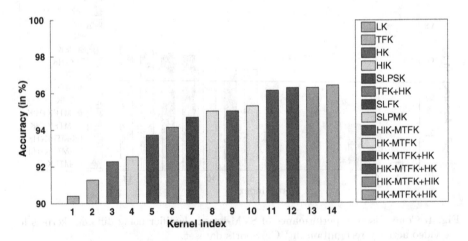

Fig. 5. Comparison of performance of SVM-based classifier using different kernels for the video activity recognition in UCF50 dataset.

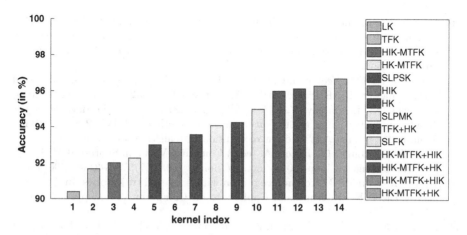

Fig. 6. Comparison of performance of SVM-based classifier using different kernels for the video activity recognition in hollywood dataset.

5 Conclusion

In this paper, we proposed approaches to SVM-based video activity recognition using different sequence kernels. A video activity comprises of a sequence of sub-activities whose time ordering is important for discriminating one video activity from the other. A sub-activity corresponds to a small segment of the video. Hence, in this work, we proposed sequence kernels that consider video segments. In this paper, we proposed modified time flexible kernel, segment level pyramid match kernel, segment level probabilistic sequence kernel and segment level Fisher kernel. The studies conducted using bench mark datasets show the effectiveness of the proposed kernels for SVM based activity recognition.

References

1. Rodriguez, M., Orrite, C., Medrano, C., Makris, D.: A time flexible kernel framework for video-based activity recognition. Image Vis. Comput. **48**, 26–36 (2016)
2. Yamato, J., Ohya, J., Ishii, K.: Recognizing human action in time-sequential images using hidden Markov model. In: IEEE Computer Society Conference on Computer Vision and Pattern Recognition (CVPR), pp. 379–385 (1992)
3. Shabou, A., LeBorgne, H.: Locality-constrained and spatially regularized coding for scene categorization. In: IEEE Conference on Computer Vision and Pattern Recognition (CVPR), pp. 3618–3625 (2012)
4. Wang, J., Liu, P., She, M., Liu, H.: Human action categorization using conditional random field. In: IEEE Workshop on Robotic Intelligence in Informationally Structured Space (RiiSS), pp. 131–135 (2011)
5. Dileep, A.D., Sekhar, C.C.: HMM based intermediate matching kernel for classification of sequential patterns of speech using support vector machines. IEEE Trans. Audio Speech Lang. Process. **21**(12), 2570–2582 (2013)

6. Sharma, N., Sharma, A., Thenkanidiyoor, V., Dileep, A.D.: Text classification using combined sparse representation classifiers and support vector machines. In: 4th International Symposium on Computational and Business Intelligence (ISCBI), pp. 181–185 (2016)
7. Van Gemert, J.C., Veenman, C.J., Smeulders, A.W.M., Geusebroek, J.-M.: Visual word ambiguity. IEEE Trans. Pattern Anal. Mach. Intell. **32**(17), 1271–1283 (2010)
8. Chatfield, K., Lempitsky, V.S., Vedaldi, A., Zisserman, A.: The devil is in the details: an evaluation of recent feature encoding methods. BMVC **2**(4), 8 (2011)
9. Soomro, K., Zamir, A.R.: Action recognition in realistic sports videos. In: Moeslund, T.B., Thomas, G., Hilton, A. (eds.) Computer Vision in Sports. ACVPR, pp. 181–208. Springer, Cham (2014). https://doi.org/10.1007/978-3-319-09396-3_9
10. Wang, H., Schmid, C.: Action recognition with improved trajectories. In: Proceedings of the IEEE International Conference on Computer Vision, pp. 3551–3558 (2013)
11. Xu, D., Chang, S.-F.: Video event recognition using kernel methods with multilevel temporal alignment. IEEE Trans. Pattern Anal. Mach. Intell. **30**(11), 1985–1997 (2008)
12. Cao, L., Mu, Y., Natsev, A., Chang, S.-F., Hua, G., Smith, J.R.: Scene aligned pooling for complex video recognition. In: Fitzgibbon, A., Lazebnik, S., Perona, P., Sato, Y., Schmid, C. (eds.) ECCV 2012. LNCS, pp. 688–701. Springer, Heidelberg (2012). https://doi.org/10.1007/978-3-642-33709-3_49
13. Vahdat, A., Cannons, K., Mori, G., Oh, S., Kim, I.: Compositional models for video event detection: a multiple kernel learning latent variable approach. In: IEEE International Conference on Computer Vision (ICCV), pp. 1185–1192 (2013)
14. Li, W., Yu, Q., Divakaran, A., Vasconcelos, N.: Dynamic pooling for complex event recognition. In: IEEE International Conference on Computer Vision (ICCV), pp. 2728–2735 (2013)
15. Shawe-Taylor, J., Cristianini, N.: Kernel Methods for Pattern Analysis. Cambridge University Press, Cambridge (2004)
16. Scovanner, P., Ali, S., Shah, M.: A 3-dimensional sift descriptor and its application to action recognition. In: Proceedings of the 15th ACM International Conference on Multimedia, pp. 357–360 (2007)
17. Klaser, A., Marszałek, M., Schmid, C.: A Spatio-temporal descriptor based on 3D-gradients. In: 19th British Machine Vision Conference (BMVC), pp. 1–275 (2008)
18. Laptev, I.: Space-time interest points. Int. J. Comput. Vis. **64**(2–3), 107–123 (2005)
19. Kliper-Gross, O., Gurovich, Y., Hassner, T., Wolf, L.: Motion interchange patterns for action recognition in unconstrained videos. In: Fitzgibbon, A., Lazebnik, S., Perona, P., Sato, Y., Schmid, C. (eds.) ECCV 2012. LNCS, vol. 7577, pp. 256–269. Springer, Heidelberg (2012). https://doi.org/10.1007/978-3-642-33783-3_19
20. Lazebnik, S., Schmid, C., Ponce, J.: Beyond bags of features: spatial pyramid matching for recognizing natural scene categories. In: IEEE Computer Society Conference on Computer Vision and Pattern Recognition (CVPR 2006), New York, vol. 2, pp. 2169–2178 (2006)
21. Gupta, S., Dileep, A.D., Thenkanidiyoor, V.: Segment-level pyramid match kernels for the classification of varying length patterns of speech using SVMs. In: 24th European Signal Processing Conference (EUSIPCO), pp. 2030–2034 (2016)
22. Zha, S., Luisier, F., Andrews, W., Srivastava, N., Salakhutdinov, R.: Exploiting image-trained CNN architectures for unconstrained video classification. arXiv preprint arXiv:1503.04144 (2015)

23. Wu, Z., Wang, X., Jiang, Y.-G., Ye, H., Xue, X.: Modeling spatial-temporal clues in a hybrid deep learning framework for video classification, pp. pp. 461–470. arXiv preprint arXiv:1504.01561 (2015)
24. Varadarajan, B., Toderici, G., Vijayanarasimhan, S., Natsev, A.: Efficient large scale video classification. arXiv preprint arXiv:1505.06250 (2015)
25. Karpathy, A., Toderici, G., Shetty, S., Leung, T., Sukthankar, R., Fei-Fei, L.: Large-scale video classification with convolutional neural networks. In: Proceedings of the IEEE conference on Computer Vision and Pattern Recognition (CVPR), pp. 1725–1732 (2014)
26. Simonyan, K., Zisserman, A.: Two-stream convolutional networks for action recognition in videos. In: Advances in Neural Information Processing Systems, pp. 568–576 (2014)
27. Wang, L., Qiao, Y., Tang, X.: Action recognition with trajectory-pooled deep-convolutional descriptors. In: Proceedings of the IEEE Conference on Computer Vision and Pattern Recognition, pp. 4305–4314 (2015)
28. Wang, L., Xiong, Y., Wang, Z., Qiao, Y.: Towards good practices for very deep two-stream ConvNets. arXiv preprint arXiv:1507.02159 (2015)
29. Wang, L., et al.: Temporal segment networks: towards good practices for deep action recognition. In: Leibe, B., Matas, J., Sebe, N., Welling, M. (eds.) ECCV 2016. LNCS, vol. 9912, pp. 20–36. Springer, Cham (2016). https://doi.org/10.1007/978-3-319-46484-8_2
30. Simonyan, K., Zisserman, A.: Very deep convolutional networks for large-scale image recognition. arXiv preprint arXiv:1409–1556 (2014)
31. Marszalek, M., Laptev, I., Schmid, C.: Actions in context. In: IEEE Conference on Computer Vision and Pattern Recognition, pp. 2929–2936 (2009)
32. Gupta, S., Thenkanidiyoor, V., Aroor Dinesh, D.: Segment-level probabilistic sequence kernel based support vector machines for classification of varying length patterns of speech. In: Hirose, A., Ozawa, S., Doya, K., Ikeda, K., Lee, M., Liu, D. (eds.) ICONIP 2016. LNCS, vol. 9950, pp. 321–328. Springer, Cham (2016). https://doi.org/10.1007/978-3-319-46681-1_39
33. Thenkanidiyoor, V., Chandra Sekhar, C.: Dynamic kernels based approaches to analysis of varying length patterns in speech and image processing tasks. In: Pattern Recognition And Big Data. World Scientific (2017)
34. Reddy, K.K., Shah, M.: Recognizing 50 human action categories of web videos. Mach. Vis. Appl. J. (MVAP) **24**, 971–981 (2012)
35. Rodriguez, M.D., Ahmed, J., Shah, M.: Action MACH: a spatio-temporal maximum average correlation height filter for action recognition. In: IEEE Conference on Computer Vision and Pattern Recognition, pp. 1–8 (2008)
36. Sharma, A., Kumar, A., Allappa, S., Thenkanidiyoor, V., Dileep, A.D.: Modified time flexible kernel for video activity recognition using support vector machines. In: 7th International Conference on Pattern Recognition Applications and Methods, pp. 133–140 (2018)

Visual Cryptography for Detecting Hidden Targets by Small-Scale Robots

Danilo Avola[1], Luigi Cinque[2], Gian Luca Foresti[1], and Daniele Pannone[2(✉)]

[1] Department of Mathematics, Computer Science and Physics, University of Udine,
Via delle Scienze 206, 33100 Udine, Italy
{danilo.avola,gianluca.foresti}@uniud.it
[2] Department of Computer Science, Sapienza University,
Via Salaria 113, 00198 Rome, Italy
{cinque,pannone}@di.uniroma1.it

Abstract. The last few years have seen a growing use of robots to replace humans in dangerous activities, such as inspections, border control, and military operations. In some application areas, as the latter, there is the need to hide strategic information, such as acquired data or relevant positions. This paper presents a vision based system to find encrypted targets in unknown environments by using small-scale robots and visual cryptography. The robots acquire a scene by a standard RGB camera and use a visual cryptography based technique to encrypt the data. The latter is subsequently sent to a server whose purpose is to decrypt and analyse it for searching target objects or tactic positions. To show the effectiveness of the proposed system, the experiments were performed by using two robots, i.e., a small-scale rover in indoor environments and a small-scale Unmanned Aerial Vehicle (UAV) in outdoor environments. Since the current literature does not contain other approaches comparable with that we propose, the obtained remarkable results and the proposed method can be considered as baseline in the area of encrypted target search by small-scale robots.

Keywords: Visual cryptography · Encrypted target
Shares generation · Target recognition · Rover · UAV · RGB camera
SLAM

1 Introduction

Over the last decade, many efforts have been made to use robots in place of humans in a wide range of complex and dangerous activities, such as search and rescue operations [1,2], land monitoring [3–5], and military missions [6,7]. In particular, the use of small-scale robots has been progressively promoted due to the following aspects: cost reduction, risk reduction for humans, and failure reduction caused by human factor (e.g., carelessness, inaccuracy, tiredness). Some significant examples are the defusing of Improvised Explosive Devices (IEDs) placed

© Springer Nature Switzerland AG 2019
M. De Marsico et al. (Eds.): ICPRAM 2018, LNCS 11351, pp. 186–201, 2019.
https://doi.org/10.1007/978-3-030-05499-1_10

on the ground and the constant monitoring of wide areas at low-altitudes. In these contexts, the use of rovers and UAVs, respectively, equipped with a vision based system can be efficiently adopted also optimizing the missions in terms of performance, speed, and security. Other examples are tasks as object recognition in outdoor environments and change detection in indoor environments, where small-scale robots are used to support moving video-surveillance systems to automatically detect novelties in the acquired video streams.

Often, especially in military field, the protection from intruders of data acquired during the exploration of areas of interest is a crucial task. This is due to the fact that the acquired data may contain sensible information, such as faces of persons, images of restricted areas, or strategic targets. To prevent the leak or the steal of information from images, in this paper a client-server based system to exploit the visual cryptography technique for encrypting acquired visual data is presented. In detail, the first step consists in using visual cryptography to generate two shares from a target image. Following a public key cryptography approach, only one share, i.e., the private key, is stored in the server. Later, the small-scale robot, used to explore the area of interest, captures the scene by an RGB camera and hides the data contained in it by using the same visual cryptography algorithm. Finally, only a share, i.e., the public key, is sent to the server and made available for subsequent decryption processes.

This paper improves and expands the work presented in [8] by introducing the following extensions:

- In addition to the small-scale rover, also a small-scale UAV is used to enrich the tests, thus confirming the goodness of the previous results;
- In addition to the single indoor environment, other challenging indoor environments (for the rover) and some challenging outdoor environments (for the UAV) are used to stress the ability of the proposed method;
- In addition to the previously adopted target objects, also areas of interest are used to test the proposed visual cryptography algorithm. In particular, parts of acquired images are used as invisible markers to identify the areas.

Like for the system presented in [8], the small-scale rover exhaustively explores the area of interest by a Simultaneous Localization and Mapping (SLAM) algorithm, while the small-scale UAV is manually piloted. This last choice is due to the fact that the implementation of a SLAM algorithm for a UAV introduces additional complex issues that, currently, are not the focus of the present paper. Notice that, as for the work presented in [8], the SLAM algorithm for the small-scale rover is inherited by the method reported in [9].

The rest of the paper is structured as follows. In Sect. 2, a brief overview about the visual cryptography is discussed. In Sect. 3, the system architecture and the visual cryptography algorithm are presented. In Sect. 4, the experimental results are shown. Finally, Sect. 5 concludes the paper.

2   Visual Cryptography Overview

As well known, the term visual cryptography is referred to a family of techniques used to encrypt an image by splitting it into n images called shares. The latter do not allow to distinguish any information about the original image unless they are combined together. This means that if only one share is available, the source data is inaccessible. In our knowledge, the only work in literature that combines small-scale robots, visual cryptography, and a SLAM algorithm (at least for the small-scale rover) is that presented in [8]. In particular, in that work, the authors proposed a client-server rover based system to search encrypted objects in unknown environments.

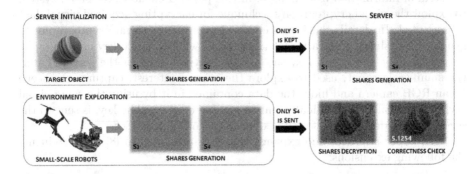

Fig. 1. System architecture. The server initialization stage shows the storing of the private key. The environment exploration stage shows the generation of the public key. Finally, the server stage shows the decryption of the data.

The visual cryptography technique was introduced by Naor and Shamir [10]. The method they proposed has never been heavily modified, but some improvements and variants can be found in the current literature [11]. Authors in [12], for example, proposed an improvement for the perfect black visual cryptography scheme, thus allowing to achieve real-time performance for the decryption step. Another example is reported in [13], where the authors proposed a verifiable multi-toned visual cryptography scheme to securely transmit confidential images (e.g., medical, forensic) on the web. Since, in visual cryptography, the image resulting from the algorithm is decrypted by the human sight, some techniques to enhance the quality of the decrypted images have been also proposed. In particular, the most popular technique is the dithering (or halftoning) [14,15], which allows to create halftone images. Some works that present a visual cryptography technique for halftone images are reported in [16–18]. In the state-of-the-art, works that combine several approaches to improve the ciphering performance [19,20] or to integrate the visual cryptography with traditional protection schemes to enhance them [21–23] can be also found.

Concerning the environment exploration, the present literature is based on the SLAM approaches [24–26]. The aim of these approaches is to maximize the area coverage during the environment exploration and, at the same time, to make the robot conscious of its absolute position within it. SLAM approaches can be used with several sensors, such as depth/time of flight cameras [27,28], thermal cameras [29], or a fusion of them [30,31]. In addition, these approaches can be used for different tasks, such as mosaicking generation [32], pipe rehabilitation [33], environment mapping [34], and others.

3 Architecture and Visual Cryptography Algorithm

In this section, the system architecture and the visual cryptography algorithm are described. Initially, the client-server approach is explained, then the steps required to encrypt and decrypt the images are reported.

3.1 System Architecture

In Fig. 1, the architecture of the proposed client-server system is shown. The small-scale rover and the small-scale UAV are considered as the client side, instead a standard workstation is considered the server side. Before starting the environment exploration, the server must be initialized. This is done by storing in it an encrypted target image T. This image can represent a specific object or a location of interest that requires to be hidden. By applying the visual cryptography algorithm on T, two shares, i.e., S_1 and S_2, are generated. The adopted public key cryptography approach is designed to store only a share, i.e., S_1, on the server. The latter is defined as the private key of the target image, while the share S_2 is discarded. The environment can be automatically explored with a small-scale rover driven by a SLAM algorithm, or manually explored by piloting a small-scale UAV. During the exploration, the used robot acquires the scene with a standard RGB camera. On the acquired images, the visual cryptography algorithm is applied to generate, once again, two shares, i.e., S_3 and S_4. While S_3 is discarded, S_4 is sent to the server and defined as public key. The latter is used in conjunction with S_1 to decrypt the target image T.

The advantage of using shares instead of clear images is that even if an intruder makes a physical attack (e.g., clients or servers are stolen) or a digital attack (e.g., a video stream is sniffed), the original information cannot be recovered. Moreover, the encryption of objects or locations of interest can be used to generate invisible markers. This is due to the fact that a target image can be decrypted only by using the correct shares. This means that when an image is decrypted, the robot is in a specific position that represents the target spot within the environment.

3.2 Visual Cryptography Algorithm

In this section, the visual cryptography algorithm is explained. For the encrypting and decrypting stages, the approach reported in [18] is used. It consists of several steps to create the shares. The first is the application of a dithering algorithm to the original image I. The dither is a form of noise intentionally applied to reduce the quantization error. As a result, I is converted into an approximate binary image so that the encryption and decryption processes are easier and the decrypted image has a good quality. In the current literature a wide range of dithering algorithms is available:

- **Average Dithering** [35]: is one of the simplest techniques. It consists in calculating the middle tone of each area and in assigning this value to that portion of image;
- **Floyd-Steinberg** [36]: is still the most used. It consists in diffusing the quantization error to the near pixels of each pixel of the image;
- **Average Ordered Dithering** [37]: is similar to average dithering, but it generates cross-hatch patterns;
- **Halftone Dithering** [36]: looks similar to newspaper halftone and produces clusters of pixel areas;
- **Jarvis Dithering** [38]: is similar to Floyd-Steinberg, but it distributes the quantization error farther than it, increasing computational cost and time.

Due to its easiness of implementation and its good quality results, in the present approach, as dithering algorithm, the Floyd-Steinberg was chosen. This algorithm diffuses the quantization error to the neighbour pixels as follows:

$$\begin{bmatrix} 0 & 0 & 0 \\ 0 & p & \frac{7}{16} \\ \frac{3}{16} & \frac{5}{16} & \frac{1}{16} \end{bmatrix} \tag{1}$$

where, the pixel p is the current pixel examined during the execution of the algorithm. Considering that I is scanned from left to right and from top to bottom, the pixels are quantized only once. Since the proposed system uses colour images, the chosen dithering algorithm is applied to each channel of I, thus obtaining three dithered images. In Fig. 2, the result of this step is shown.

    (a)   (b)   (c)

Fig. 2. Dithered images for channels: (a) cyano, (b) magenta, and (c) yellow. (Color figure online)

(a)   **(b)**

Fig. 3. Sharing and stacking combination in grayscale images. In both pictures, (a) and (b), the first column is the original pixel (i.e., black and white, respectively), the second and third columns are the *share 1* and *share 2*, respectively. Finally, the last column is the stacked shares.

Algorithm 1. Algorithm generating shares from a grayscale image.

1: **procedure** SHARESFROMGRAY(I)
2:     Transform I to halftone image H
3:     **for** each pixel in H **do**
4:         Choose randomly a share among those in
5:             Figure 3
6:     **end for**
7: **end procedure**

Algorithm 2. Algorithm generating shares from a color image.

1: **procedure** SHARESFROMCOLOR(colorImage)
2:     Transform colorImage in three halftone images C, M and Y
3:     **for** each pixel $p_{i,j}$ in C,M and Y **do**
4:         According to Algorithm 1, create
5:             $C1_{i,j}, C2_{i,j}, M1_{i,j}, M2_{i,j}, Y1_{i,j}, Y2_{i,j}$
6:         Combine $C1_{i,j}, M1_{i,j}$ and $Y1_{i,j}$ for the corresponding block of Share 1
7:         Combine $C2_{i,j}, M2_{i,j}$ and $Y2_{i,j}$ for the corresponding block of Share 2
8:     **end for**
9:     After stacking the two Shares, the original image can be decrypted.
10: **end procedure**

After the generation of the dithered images, also the shares can be created. Since the latter are generated starting from grayscale images, the shares generation algorithm can be defined as follows:

1. I is tranformed into a black and white halftone image H;
2. For each pixel in the halftone image, a random combination is chosen among those depicted in Fig. 3;
3. Repeat step 2 until every pixel in H is decomposed.

The pseudo-code is reported in Algorithm 1. To generate the shares for colour images, the third method presented in [18] was used. The choice fell on this method since it requires only two shares to encrypt/decrypt a colour image, in addition, it does not sacrifice too much image contrast in the resulting image. The method works as follows. First, a dithered image for each channel of I is created. Assuming that we are using the YCMK profile, we obtain a dithered

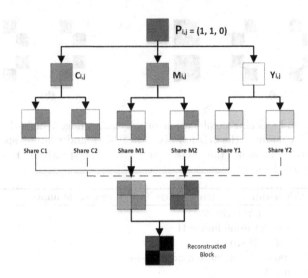

Fig. 4. Decomposition and reconstruction of colour pixel [8].

image for *Cyan* (C), *Magenta* (M), and *Yellow* (Y) channels. Subsequently, for each halftone image the Algorithm 1 is used to generate six 2×2 sharing images, called *C1, C2, M1, M2, Y1*, and *Y2*. Each of these shares is composed by two white pixels and two colour pixels. To generate the coloured *share 1, C1, M1*, and *Y1* are combined together, while for generating the coloured *share 2, C2, M2*, and *Y2* are combined. The colour intensity of the share blocks is a value between 0 and 1, where 0 means the absence of that colour and 1 means full intensity. So, for a pixel $p_{i,j}$ the colour intensity for each channel is defined as (I_C, I_M, I_Y). For each block generated with this method, we have that the colour intensity is $(\frac{1}{2}, \frac{1}{2}, \frac{1}{2})$, while after stacking *shares 1* and *shares 2*, the range of colour intensity is between $(\frac{1}{2}, \frac{1}{2}, \frac{1}{2})$ and $(1, 1, 1)$. As for the grayscale algorithm, the decryption step simply consists in overlapping the two shares, thus obtaining the decrypted image I_{dec}. In Algorithm 2 the pseudo-code of the method is shown, while in Fig. 4 a representation of the algorithm is depicted.

Since the images acquired by the robot may not be acquired at the same distance, position, and angulation of I, and since the pixels composing the two shares must be almost perfectly aligned to perform the decryption, a morphing procedure is applied on S_4. This allows to optimize, in some cases, the alignment of the two shares [39,40]. Considering that with the shares a feature-based (e.g., by using keypoints and homography) alignment cannot be performed due to their random pixel arrangement, we have defined eight standard transformations to apply. In Fig. 5, these transformations are shown.

After the generation of I_{dec}, a check to verify the validity of the decrypted image is performed. In particular, the difference between the standard deviation of I_{dec} and its smoothed copy $Smooth_{I_{dec}}$ is computed. The smoothing operation is performed by using a median filter with kernel size of 3×3. Formally:

$$Correctness = I_{dec} - Smooth_{I_{dec}} \tag{2}$$

If I_{dec} is a valid decrypted image, we have that the standard deviation between it and its smoothed copy is low (e.g., less than 10), otherwise higher values are obtained.

3.3 SLAM Algorithm

This section briefly reports the SLAM algorithm, previously presented in [9], used by the rover to exhaustively analyse an unknown environment. Notice that, the used rover is composed by two main components. The first is the base, which is composed in turn by the tracks, while the second is the upper part, that contains the RGB sensor and the ultrasonic sensor. Between the base and the upper part a servomotor is mounted, in a way such that the two parts can be moved independently from each other.

The algorithm considers the environment to be explored as a two-dimensional Cartesian plane $C(X, Y)$, composed by points of the form $c(x, y)$ reachable by the rover. At each c, the rover examines the environment by rotating the base of $0°, 90°, 180°$, and $270°$ with respect to the x axis. Once the rover has assumed a position, it starts moving the upper part at $60°, 90°$, and $120°$ with respect to its local coordinates, thus acquiring an image for each of these angles. If T is not found in these images, the rover checks with the proximity sensor the next point c in which it can move. Once obtained the distance d with the proximity sensor, the robot checks if it is less than the distance $d_{threshold}$, which represents the maximum distance within which the sensor detects an obstacle. The threshold depends on the size of the robot, the size of the object to be searched, and the resolution of the images acquired by the RGB sensor. If $d < d_{threshold}$, it means that there is an obstacle and the rover cannot move to that point. Otherwise, the rover moves to the new point c' and repeat the steps.

With respect to the work presented in [8], an improvement that has been made to the SLAM algorithm is the addition of the tilt movement of the RGB sensor. This allows enhancing the acquisition of flat objects, such as credit cards, books, and so on. The tilt movement is performed by moving the camera of $30°, 60°$, and $75°$ with respect to the camera starting position.

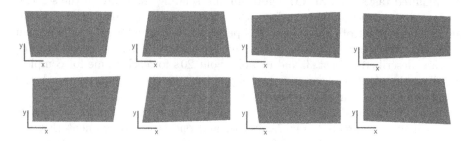

Fig. 5. Transformations applied to S_4 for optimizing the shares overlap.

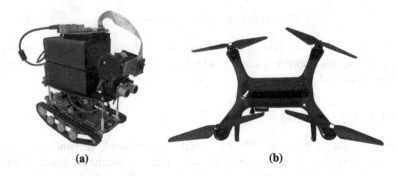

Fig. 6. The used robots: (a) the small-scale rover, (b) the small-scale UAV.

4   Experimental Results

In this section, the experimental tests are presented. In detail, we first report the experiments performed with the small-scale rover (Fig. 6a), then the experiments performed with the small-scale UAV (Fig. 6b). Considering that, currently, there are no datasets for this kind of systems/methods, in all the experiments our acquisitions are used. The experiments were performed in controlled conditions, with unnoticeable changes of illumination, and without moving objects. The experiments with the small-scale rover were performed in three indoor environments, while the experiments with the small-scale UAV were performed in a wide outdoor environment. The communication between robots and server was made by a direct Wi-Fi connection, to reduce the delay introduced by sending the network packets.

4.1   Experiments with the Small-Scale Rover

In Fig. 6a, the small-scale rover used in three indoor environments is shown. It is composed by an Arduino UNO micro-controller, which handles both the servomotors and the ultrasonic sensor used by the SLAM algorithm, and by a Raspberry Pi 2 model B, which handles the camera/video stream and the communications with the server. Despite the used rover is able to perform all the required tasks (i.e., SLAM algorithm and sending the share to the server), its low computational power affects the time needed to explore the environment. The first set of experiments was performed by running the entire system on-board. After capturing the frame, the Raspberry Pi 2 proceeded with the pipeline described in Sect. 3, and it took about 20 s for each frame for completing the entire pipeline. To overcome this problem, the client-server approach was adopted for improving the performance.

In Fig. 7, the used indoor environments are shown. In detail, we chosen two challenging environments (Fig. 7a and b) and one easy environment (Fig. 7c) to test exhaustively both the SLAM algorithm and the visual cryptography pipeline. The challenging environments are rooms containing desks and seats, while the easy environment is a hallway. For each environment, the rover starts the recognition always from the same starting point. To stress the system, we used both clear (i.e., just placed on the ground) and covered (i.e., underneath the desks or the seats) objects. A total of 20 objects (reported in Table 1) with high variability of colours and sizes was used.

Table 1. List of objects used during the experiments [8].

Object number	Object type	Object type	Object type
Object 1	Ball	Object 11	Cup
Object 2	Toy Gun	Object 12	Coffe Machine
Object 3	USB Keyboard	Object 13	Wallet
Object 4	Pen	Object 14	Plastic Bottle
Object 5	Calculator	Object 15	Monitor
Object 6	Pencil	Object 16	Toy Robot
Object 7	Paperweight	Object 17	DVD Case
Object 8	Credit card	Object 18	Small Box
Object 9	Sponge	Object 19	Big Box
Object 10	USB mouse	Object 20	Book

Concerning the first challenging environment (Fig. 7a), the confusion matrix is reported in Table 2. In this environment, we have noticed that a good correctness value is in the range of $[4, 6]$, while a wrong decrypted image has a correctness value between $[19, 25]$. As it is possible to see, the proposed system works generally well, but due to their characteristics the decryption fails for the credit card and the cup. In detail, the decryption fails because even if we apply the transformation shown in Fig. 5, it may be not sufficient to correctly align the shares. The sponge, differently from the experiments presented in [8], was correctly decrypted thanks to the tilt movement of the RGB camera implemented in this new version of the SLAM algorithm.

Regarding the second environment, the correctness values are reported in Table 3. These values slightly differ from the ones obtained in the first indoor environment due to the different illumination conditions. In this case, the values obtained for good decrypted images are in the range of $[7, 9]$, while the values corresponding to bad decrypted images are in the range of $[27, 31]$. As for the first environment, the credit card is hard to decrypt, due to its flat shape that makes difficult to acquire correctly the object and then to generate a good share.

Finally, in Table 4 the correctness values of the third indoor environment are shown. As for the previous environments, we have low values for a correct decrypted image and high values otherwise. In detail, correct values are in the range of $[3, 5]$, while high values are in the range of $[15, 18]$. Since this is an easy environment, we used only a subset of the objects, and to make the experiments

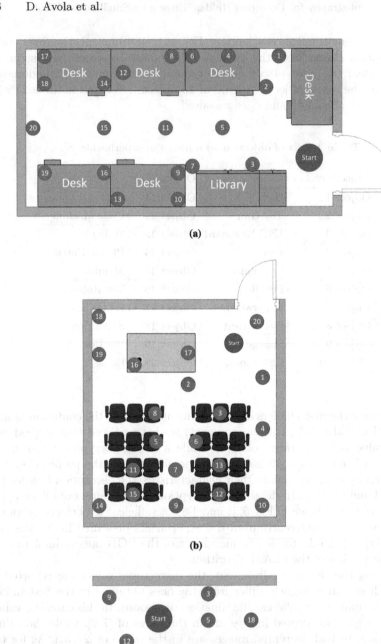

Fig. 7. Environments used for indoor experiments: (a) and (b) challenging environments, (c) an easy (c) environment. Each environment was chosen to test the system in different conditions. The dark blue circles are the objects placed in clear, while the dashed green circles represent the covered objects. (Color figure online)

Table 2. Confusion matrix of the first indoor environment [8].

	Object 1	Object 2	Object 3	Object 4	Object 5	Object 6	Object 7	Object 8	Object 9	Object 10	Object 11	Object 12	Object 13	Object 14	Object 15	Object 16	Object 17	Object 18	Object 19	Object 20
Object 1	4.9792	24.2623	21.1189	21.6967	24.7812	19.2538	24.8378	20.1352	23.0027	22.5186	23.0507	21.1661	22.7217	23.8669	19.1155	19.5032	24.8488	22.9081	20.3874	21.4209
Object 2	19.7321	4.5369	20.5471	20.9900	19.9134	21.0880	19.3700	24.3049	19.5657	24.5802	21.3941	19.2844	21.0542	23.4158	23.7681	22.2694	23.1173	24.3418	19.3288	20.8220
Object 3	19.2771	20.1729	5.4403	23.3305	24.2668	22.4946	19.4241	24.5365	23.8022	20.7157	22.2620	24.9087	23.2941	24.0338	21.5996	21.8237	22.3643	20.6145	23.4941	22.0233
Object 4	22.8809	20.8465	19.8323	4.9511	21.1748	23.7287	23.6818	23.0111	19.8010	19.1293	22.3590	20.8049	24.6365	24.4854	20.7197	23.8049	24.3767	22.5852	24.3041	24.6624
Object 5	22.2949	23.3703	22.4605	19.1551	4.8931	22.8778	22.1272	21.2339	24.6228	23.9772	24.0945	21.2352	22.5591	24.2353	24.6010	23.0108	20.2407	22.9231	19.4323	21.4404
Object 6	23.0016	24.6024	23.8657	21.9073	23.5405	4.8341	24.8307	24.9278	24.1849	21.3333	21.7285	20.4801	23.7065	24.2970	24.4823	22.3497	22.5932	19.8933	24.3983	21.7024
Object 7	20.2340	24.3979	23.5755	24.2949	20.7097	23.0394	5.3286	19.7369	21.4439	20.6517	23.3000	20.7003	24.3772	23.9595	21.3402	21.9874	23.1688	24.0062	22.6578	22.4484
Object 8	20.9563	21.7385	23.2828	24.3064	23.3251	19.1117	23.0487	18.7589	21.6269	19.7022	23.8881	20.9491	20.4774	21.0563	21.2542	22.2793	22.3715	21.3749	21.3888	22.0922
Object 9	22.9452	24.7055	23.3341	21.4005	23.9912	19.8060	19.3628	19.5055	5.1504	20.9453	20.8104	19.0701	22.2394	19.5722	19.8791	22.7868	24.1559	24.8453	22.4250	24.9811
Object 10	22.3212	22.0928	20.9841	21.5800	21.9508	19.4262	24.3264	19.3878	21.6171	5.6533	21.3672	22.6808	23.9118	24.3174	24.5867	20.1447	20.5515	24.3872	22.5602	22.0230
Object 11	22.6769	23.9165	22.1913	20.2125	21.7234	21.5675	24.7963	22.7303	21.1723	23.3510	14.3278	22.1019	22.9390	23.7237	23.1688	21.5587	24.0176	23.3883	21.1602	
Object 12	21.7253	21.3183	23.6553	23.4056	21.5817	23.1625	24.6713	23.7054	23.2334	19.6560	21.3396	1.8818	22.7563	19.3020	20.3721	24.0651	19.0699	24.1823	19.4684	23.0143
Object 13	22.0013	20.3080	22.4297	19.7331	23.0270	22.5975	19.3359	19.3381	19.9150	19.1177	21.6111	23.9933	5.2349	22.1208	24.1832	19.5862	24.4483	19.6481	22.1020	19.8589
Object 14	22.3562	19.0275	23.6001	24.0923	24.5009	24.9218	22.0308	20.6285	19.6045	22.0471	22.5137	23.5773	19.4978	5.3232	22.1019	20.0263	24.6313	22.5429	21.6438	24.6515
Object 15	22.9355	21.7117	24.0382	22.1957	22.3233	23.0804	21.2031	20.4357	22.4735	24.2013	21.4407	19.6757	21.6631	20.8011	4.8028	24.0002	21.4218	21.3411	21.1627	19.8415
Object 16	20.5608	19.5209	21.5764	20.5437	20.7853	21.5492	19.7152	21.9704	23.2384	20.4614	23.7104	19.4445	21.3633	19.0204	20.3241	4.0026	20.1351	19.8549	20.6085	20.0494
Object 17	19.8319	22.5933	24.4063	24.6363	20.3271	21.8960	21.2561	22.1427	20.5892	19.4101	21.6180	20.0431	19.1566	24.7281	21.5836	24.7694	5.5248	19.0441	23.0802	23.2357
Object 18	22.8708	22.3139	20.3087	23.6342	20.3682	21.2252	24.3456	24.1383	21.4146	20.9081	22.6518	24.4612	24.4546	22.5496	20.9954	24.1184	21.6544	5.8087	19.1991	22.1946
Object 19	23.2990	20.0758	21.0192	20.1263	20.9316	21.4231	22.2914	19.2924	23.2164	20.6489	20.4490	24.4589	19.9250	24.7385	24.6140	23.9123	23.3696	20.0549	4.7207	20.1327
Object 20	19.0072	20.8985	23.1977	22.7515	22.2584	21.6342	20.7246	22.0100	23.5693	23.5744	24.2563	23.4860	22.8732	19.7393	22.0264	21.0836	19.5529	19.8871	20.1890	19.8832

Table 3. Confusion matrix of the second indoor environment.

	Object 1	Object 2	Object 3	Object 4	Object 5	Object 6	Object 7	Object 8	Object 9	Object 10	Object 11	Object 12	Object 13	Object 14	Object 15	Object 16	Object 17	Object 18	Object 19	Object 20
Object 1	8.6294	30.6232	27.5079	30.6535	29.5294	27.3902	26.1140	29.1875	30.8300	30.8596	27.6305	30.8824	30.8287	28.9415	30.2011	27.5675	28.6870	30.6629	30.1688	30.8380
Object 2	29.6230	7.0714	30.3965	30.7360	29.7149	30.0310	29.9725	28.5689	29.6219	27.6847	29.8242	27.1273	28.1077	27.1847	27.3885	30.2938	29.7793	28.2684	30.8009	27.1378
Object 3	28.7550	28.5262	6.5310	30.1808	27.7475	28.5591	27.5853	29.8375	30.0187	28.1041	29.7188	29.6204	26.6504	27.4760	28.9935	30.8390	28.3615	29.3411	27.8952	
Object 4	30.0051	28.0204	29.0238	8.3982	30.5636	30.8372	29.1889	27.5545	27.5972	28.0500	30.3629	28.0171	30.2571	27.9741	30.7171	28.3999	27.7864	28.0043	29.4662	28.8932
Object 5	28.4066	30.3233	29.3411	29.1989	8.8344	28.1434	30.0288	30.0149	28.5218	29.2713	27.9034	27.2158	29.1232	30.1167	30.7360	27.5196	29.2753	28.8776	27.0476	28.3485
Object 6	27.6487	30.1771	28.2449	29.1141	27.6626	8.2040	28.0519	29.9163	29.7569	29.9926	28.8022	27.3353	27.9159	30.6533	27.6095	30.3033	29.1534	30.9845	27.3127	28.7707
Object 7	27.4266	30.8476	27.0185	30.0996	30.2692	30.4748	7.1689	28.5991	28.0395	30.2003	27.7257	30.6426	27.7274	28.0552	27.5822	27.5443	30.4772	29.3188	29.1994	27.5798
Object 8	30.4121	29.4882	28.4038	29.0530	28.6072	27.3039	27.9597	7.2466	27.7356	27.9598	28.8691	27.1986	30.6109	30.7791	28.9635	28.9570	28.3509	30.6002	28.4770	27.4448
Object 9	30.1210	28.5590	27.9668	28.6156	27.3858	27.5279	30.7682	30.8245	7.4155	28.2500	27.9391	28.4126	30.0487	27.6016	27.1721	27.6760	29.5965	29.9269	29.5910	28.8037
Object 10	29.1880	28.1853	29.9788	27.7558	29.7471	27.7340	28.4739	29.5025	30.1209	7.1623	30.0175	30.1029	28.9472	28.7434	28.7871	28.2254	29.0340	29.0431	30.2705	30.1793
Object 11	29.5773	28.5144	30.2463	29.1313	28.4029	30.7560	30.5038	29.2006	29.4899	29.3482	7.4155	28.2050	28.8837	29.7220	30.3772	27.7791	27.9037	27.6828	27.9107	28.7428
Object 12	28.2444	30.6935	28.7208	27.7393	30.6195	30.9190	28.7555	27.4445	28.0823	28.6349	29.3796	7.5244	29.4114	29.8449	27.8870	27.4697	28.1867	28.2751	28.6967	29.0334
Object 13	27.3421	28.4690	29.2038	30.7154	29.9213	28.9544	29.3141	27.9491	28.8354	30.8524	29.1789	8.0423	27.9264	28.9556	29.4962	29.7165	28.5821	28.4697	30.9519	
Object 14	27.1510	30.5407	30.6531	30.1847	27.3948	28.0475	28.3414	29.7189	27.5462	29.8849	27.4270	29.6150	28.9767	8.5581	29.8601	30.6149	30.5637	28.3367	29.7950	27.7912
Object 15	27.1220	29.9763	29.0001	28.2197	30.6189	29.4395	29.4707	30.4378	30.2220	29.5069	27.7317	27.9597	30.5460	27.1147	7.9376	30.2013	28.7296	30.3013	27.3339	27.5327
Object 16	27.2385	29.7729	27.1697	27.2858	29.0866	27.3869	30.2702	29.8898	27.5999	29.6384	29.0744	30.8919	29.5960	30.2013	7.9976	28.7296	30.9147	28.4501	29.0019	28.8844
Object 17	27.6936	28.5638	30.3255	30.2135	27.2419	28.5970	29.1075	28.6672	29.6274	29.5119	28.1679	28.7266	27.0619	30.9563	27.6687	27.4249	7.7448	27.7925	28.9588	28.3580
Object 18	30.8065	30.6813	27.2107	29.9514	28.0765	28.6913	29.1915	30.7709	28.6710	30.9322	28.2058	29.8044	26.9255	29.5269	29.6661	27.1123	29.5010	8.0540	30.9963	27.6845
Object 19	27.1304	29.2448	30.5275	29.6767	27.7617	28.4757	28.8429	30.9266	27.6256	30.4221	29.5791	28.5051	27.7637	28.7130	28.9281	27.4824	29.3580	27.9048	7.7692	29.3319
Object 20	28.0072	28.1618	29.4684	28.0611	30.2975	30.9307	29.9210	28.7355	29.3363	27.4311	30.6252	30.5186	30.2710	28.0429	29.3774	27.0901	28.7010	28.2509	27.6459	7.3575

Table 4. Confusion matrix of the third indoor environment.

	Object 3	Object 5	Object 9	Object 12	Object 18
Object 3	3.5521	17.0391	16.9653	15.4878	15.3570
Object 5	16.4951	4.9195	16.0212	16.7558	15.6714
Object 9	17.2538	15.7653	4.0119	17.0972	17.6727
Object 12	17.8779	16.6416	15.4159	3.2986	15.7725
Object 18	17.5222	15.7628	17.4429	15.7306	4.8585

more challenging, we tried to occlude the targets. Despite the uncovered parts were correctly decrypted, we obtained high values for the correctness measures due to the occluded ones.

Regarding the execution time, to explore the first and second environment the rover took about 20 to 50 min for each one. This is due to the random approach of the used SLAM algorithm, and also due to the fact that for this kind of rover the exploration under the desks is difficult. Comparing with the timing reported in [8], the lower bound has been improved by 10 min. This is due to the tilt movement implemented in the new version of the SLAM algorithm, which allows to find targets quickly. Concerning the third environment, the rover took about 10 to 25 min to explore it. This is mainly due to the fact that no obstacles, such as seats and desks, were present.

4.2 Experiments with the Small-Scale UAV

In Fig. 6b, the small-scale UAV used in a wide outdoor environment is shown. The rotors are handled by a Pixhawk 2, while the camera, the visual cryptography algorithm, and the Wi-Fi connection are managed, as for the rover, by the Raspberry Pi 2. Concerning the correctness values, they are reported in Table 5. In these experiments, we obtained the best results in term of correctness. This is due to the fact that the pose of the object placed on the ground is the same to the pose of the object at the time of the server initialization. By using the UAV, also flat objects, such as credit cards, can be correctly decrypted (as depicted in Fig. 8). A condition that must be respected for achieving a correct decryption is that the drone must perform a stabilized flight. This means that the flight height must be the same both during the target acquired and during the server initialization. In fact, zoom in/out activities (i.e., change in flight height), pitch, roll, and jaw movements can influence the share alignment.

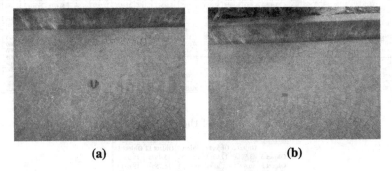

(a)   (b)

Fig. 8. The cup (a) and the credit card (b) viewed from the UAV.

Table 5. Confusion matrix of the outdoor environment.

	Object 1	Object 2	Object 3	Object 4	Object 5	Object 6	Object 7	Object 8	Object 9	Object 10	Object 11	Object 12	Object 13	Object 14	Object 15	Object 16	Object 17	Object 18	Object 19	Object 20
Object 1	3.3147	20.7174	18.3810	19.8971	20.7401	18.2926	18.8355	19.6406	20.8725	20.8947	18.4728	20.9118	20.8715	19.4561	20.4008	18.4257	19.2653	20.7472	20.3766	20.8785
Object 2	19.9672	2.5357	20.5474	20.8020	20.0362	20.2732	20.2294	19.1767	19.9664	18.5136	20.1181	18.0955	18.8308	18.1385	18.2914	20.4704	20.0845	18.9513	20.8507	18.1033
Object 3	19.3162	19.1447	3.2655	20.3856	18.5606	19.4693	19.3368	19.9389	20.1281	20.2641	18.8281	20.0391	19.9653	18.4878	18.3570	19.4951	20.8792	19.0212	19.7558	18.6714
Object 4	20.2538	18.7653	19.5179	3.1991	20.6727	20.8779	19.6416	18.4159	18.4479	18.7725	20.5222	18.7628	20.4429	18.7306	20.7878	19.0500	18.5898	18.7533	19.8481	19.4199
Object 5	19.0550	20.4925	19.7558	19.6492	3.4172	18.8575	20.2716	20.2612	19.1413	19.7035	18.2276	18.1619	19.3924	20.3375	20.8020	18.3897	19.7065	19.4082	18.0357	19.0114
Object 6	18.4865	20.3829	18.9336	19.5856	18.4969	3.1020	18.7889	19.9622	20.0676	20.2445	19.3516	18.2515	18.6869	20.7400	18.4571	20.4775	19.6150	20.9884	18.2345	19.3280
Object 7	18.3200	20.8857	18.0139	20.3247	20.4519	20.6061	2.5844	19.1993	18.7796	20.4002	19.2942	20.7319	18.5455	18.7914	18.4366	18.4082	20.6079	19.7391	19.6496	18.4349
Object 8	20.5591	19.8662	20.0529	19.5397	19.2054	18.2279	18.7197	2.6233	18.5517	18.7199	19.2518	18.1490	20.7081	20.8344	19.4726	19.4678	19.0132	20.7002	19.1077	18.3336
Object 9	20.3408	19.1692	18.7251	19.2117	18.2894	18.3959	20.8262	20.8684	3.0752	18.1793	18.7043	19.0595	20.4636	18.0462	18.1291	18.5070	19.9473	20.1952	19.9432	19.3528
Object 10	19.6410	18.8890	20.2341	18.5669	20.0603	18.5505	19.1055	19.8769	20.3407	2.5811	20.7882	20.3271	19.4604	19.3076	19.3404	18.9190	19.5255	19.5323	20.4529	20.3845
Object 11	19.9330	19.1358	20.4347	19.5985	19.0522	20.8170	20.6278	19.6505	19.8674	19.7611	2.7077	18.9037	19.4128	18.6915	20.5329	18.5843	18.6778	18.5121	18.6830	19.3071
Object 12	18.9333	20.7701	19.2906	18.5544	20.7146	20.9392	19.3166	18.3334	18.7742	19.2262	19.7847	2.7622	19.8085	20.1336	18.6652	18.3523	18.8900	18.9563	19.2725	19.5236
Object 13	18.2565	18.7874	20.4030	18.0877	20.7866	20.1910	19.4658	19.7356	18.7119	19.3765	20.8893	19.6404	3.0211	18.6948	19.4667	19.8722	20.0374	19.1865	19.1023	20.9639
Object 14	18.1132	20.6555	20.7399	20.3886	18.2961	18.7856	19.0061	20.0392	18.4097	20.1637	18.3203	19.9613	19.4825	3.2791	20.1451	20.7112	20.6728	19.0025	20.0962	18.5934
Object 15	18.0916	20.2322	19.5001	19.4398	20.7142	19.8296	19.8530	20.5783	20.4165	19.7302	18.5488	18.7198	20.6595	18.0860	2.9899	18.5038	20.9360	20.1381	19.5014	19.4133
Object 16	18.1789	20.5449	18.1273	18.2143	19.5649	18.2902	20.4544	20.4526	20.1673	18.4496	19.9788	19.5558	20.9189	19.9470	20.4010	2.9538	19.2972	20.4759	18.2504	18.3995
Object 17	18.5202	19.1728	20.4941	20.4101	18.1814	19.1978	19.5406	19.2504	19.9706	19.8839	18.8760	19.2950	18.0465	20.9522	18.5015	18.3186	2.8726	18.5944	19.4691	19.0185
Object 18	20.8549	20.7610	18.1580	20.2136	18.8074	19.2685	19.6436	20.8282	19.2532	20.0492	18.9044	20.1033	19.9990	19.6174	20.0943	19.9996	18.5344	2.6280	20.9972	18.5134
Object 19	18.0978	19.6836	20.6456	20.0075	18.5713	19.1067	19.3822	20.9449	18.4692	20.5666	19.9343	19.1288	18.5728	19.2848	19.4461	18.3618	19.7685	18.6786	2.8846	19.7490
Object 20	18.7554	18.8713	19.8513	18.7959	20.4731	20.9480	20.1907	19.0316	19.7522	18.3233	20.7189	20.6390	20.4533	18.7822	19.7831	18.0675	19.2758	19.9382	18.4845	2.6788

5 Conclusions

In recent years, autonomous (or semi-autonomous) small-scale robots have been increasingly used to face dangerous activities, including civilian and military operations. Usually, these robots send the acquired data to a ground station to perform a wide range of processing. In some cases, there may be the need to protect the sent data from intruders. In this paper, a system to encrypt video streams acquired by small-scale robots engaged in exploration tasks is presented. In particular, the paper shows results for two small-scale robots: a rover and a UAV. While the latter is manually piloted during the mission, the first is equipped with a SLAM algorithm that allows it to explore autonomously the environment and search exhaustively different targets. The experimental tests where performed in indoor and outdoor environments showing the effectiveness of the proposed method. Currently, in literature, there are no other approaches comparable with that we propose. For this reason, it can be considered as baseline in the area of encrypted target search by small-scale robots.

References

1. Cacace, J., Finzi, A., Lippiello, V., Furci, M., Mimmo, N., Marconi, L.: A control architecture for multiple drones operated via multimodal interaction in search rescue mission. In: IEEE International Symposium on Safety, Security, and Rescue Robotics, pp. 233–239 (2016)
2. Kiyani, M.N., Khan, M.U.M.: A prototype of search and rescue robot. In: International Conference on Robotics and Artificial Intelligence, pp. 208–213 (2016)
3. Avola, D., Foresti, G.L., Martinel, N., Micheloni, C., Pannone, D., Piciarelli, C.: Real-time incremental and geo-referenced mosaicking by small-scale UAVs. In: Battiato, S., Gallo, G., Schettini, R., Stanco, F. (eds.) ICIAP 2017. LNCS, vol. 10484, pp. 694–705. Springer, Cham (2017). https://doi.org/10.1007/978-3-319-68560-1_62
4. Avola, D., Foresti, G.L., Martinel, N., Micheloni, C., Pannone, D., Piciarelli, C.: Aerial video surveillance system for small-scale UAV environment monitoring. In: IEEE International Conference on Advanced Video and Signal Based Surveillance, pp. 1–6 (2017)
5. Avola, D., Cinque, L., Foresti, G.L., Martinel, N., Pannone, D., Piciarelli, C.: A UAV video dataset for mosaicking and change detection from low-altitude flights. IEEE Trans. Syst. Man Cybern. Syst. **PP**, 1–11 (2018)
6. Kaur, T., Kumar, D.: Wireless multifunctional robot for military applications. In: International Conference on Recent Advances in Engineering Computational Sciences, pp. 1–5 (2015)
7. Kopulet´y, M., Palasiewicz, T.: Advanced military robots supporting engineer reconnaissance in military operations. In: Mazal, J. (ed.) MESAS 2017. LNCS, vol. 10756, pp. 285–302. Springer, Cham (2018). https://doi.org/10.1007/978-3-319-76072-8_20
8. Avola, D., Cinque, L., Foresti, G.L., Marini, M.R., Pannone, D.: A rover-based system for searching encrypted targets in unknown environments. In: International Conference on Pattern Recognition Applications and Methods, vol. 1, pp. 254–261 (2018)

9. Avola, D., Foresti, G.L., Cinque, L., Massaroni, C., Vitale, G., Lombardi, L.: A multipurpose autonomous robot for target recognition in unknown environments. In: IEEE International Conference on Industrial Informatics, pp. 766–771 (2016)

10. Naor, M., Shamir, A.: Visual cryptography. In: De Santis, A. (ed.) EUROCRYPT 1994. LNCS, vol. 950, pp. 1–12. Springer, Heidelberg (1995). https://doi.org/10.1007/BFb0053419

11. Liu, S., Fujiyoshi, M., Kiya, H.: A cheat preventing method with efficient pixel expansion for Naor-Shamir's visual cryptography. In: IEEE International Conference on Image Processing, pp. 5527–5531 (2014)

12. Li, P., Yang, C.N., Kong, Q.: A novel two-in-one image secret sharing scheme based on perfect black visual cryptography. J. R. Time Image Process. **14**, 41–50 (2018)

13. Shivani, S.: VMVC: verifiable multi-tone visual cryptography. Multimed. Tools Appl. **77**, 5169–5188 (2018)

14. Alex, N.S., Anbarasi, L.J.: Enhanced image secret sharing via error diffusion in halftone visual cryptography. In: International Conference on Electronics Computer Technology, pp. 393–397 (2011)

15. Pahuja, S., Kasana, S.S.: Halftone visual cryptography for color images. In: International Conference on Computer, Communications and Electronics, pp. 281–285 (2017)

16. Lin, C.C., Tsai, W.H.: Visual cryptography for gray-level images by dithering techniques. Pattern Recognit. Lett. **24**, 349–358 (2003)

17. Babu, R., Sridhar, M., Babu, B.R.: Information hiding in gray scale images using pseudo-randomized visual cryptography algorithm for visual information security. In: International Conference on Information Systems and Computer Networks, pp. 195–199 (2013)

18. Hou, Y.C.: Visual cryptography for color images. Pattern Recognit. **36**, 1619–1629 (2003)

19. Stinson, D.R.: An introduction to visual cryptography. In: Public Key Solutions, pp. 28–30 (1997)

20. Shyu, S.J.: Efficient visual secret sharing scheme for color images. Pattern Recognit. **39**, 866–880 (2006)

21. Yang, D., Doh, I., Chae, K.: Enhanced password processing scheme based on visual cryptography and OCR. In: International Conference on Information Networking, pp. 254–258 (2017)

22. Kadhim, A., Mohamed, R.M.: Visual cryptography for image depend on RSA algamal algorithms. In: Al-Sadeq International Conference on Multidisciplinary in IT and Communication Science and Applications, pp. 1–6 (2016)

23. Joseph, S.K., Ramesh, R.: Random grid based visual cryptography using a common share. In: International Conference on Computing and Network Communications, pp. 656–662 (2015)

24. Leonard, J.J., Durrant-Whyte, H.F.: Simultaneous map building and localization for an autonomous mobile robot. In: IEEE/RSJ International Workshop on Intelligent Robots and Systems, Intelligence for Mechanical Systems, pp. 1442–1447 (1991)

25. Sim, R., Roy, N.: Global a-optimal robot exploration in slam. In: IEEE International Conference on Robotics and Automation, pp. 661–666 (2005)

26. Trivun, D., Šalaka, E., Osmanković, D., Velagić, J., Osmić, N.: Active SLAM-based algorithm for autonomous exploration with mobile robot. In: IEEE International Conference on Industrial Technology, pp. 74–79 (2015)

27. Li, C., Wei, H., Lan, T.: Research and implementation of 3D SLAM algorithm based on kinect depth sensor. In: International Congress on Image and Signal Processing, BioMedical Engineering and Informatics, pp. 1070–1074 (2016)
28. Walas, K., Nowicki, M., Ferstl, D., Skrzypczyński, P.: Depth data fusion for simultaneous localization and mapping - RGB-DD SLAM. In: IEEE International Conference on Multisensor Fusion and Integration for Intelligent Systems, pp. 9–14 (2016)
29. Chen, L., Sun, L., Yang, T., Fan, L., Huang, K., Xuanyuan, Z.: RGB-T SLAM: a flexible slam framework by combining appearance and thermal information. In: IEEE International Conference on Robotics and Automation, pp. 5682–5687 (2017)
30. Mur-Artal, R., Tardós, J.D.: ORB-SLAM2: an open-source slam system for monocular, stereo, and RGB-D cameras. IEEE Trans. Robot. **33**, 1255–1262 (2017)
31. Camurri, M., Bazeille, S., Caldwell, D.G., Semini, C.: Real-time depth and inertial fusion for local SLAM on dynamic legged robots. In: IEEE International Conference on Multisensor Fusion and Integration for Intelligent Systems, pp. 259–264 (2015)
32. Bu, S., Zhao, Y., Wan, G., Liu, Z.: Map2DFusion: real-time incremental UAV image mosaicing based on monocular slam. In: IEEE/RSJ International Conference on Intelligent Robots and Systems, pp. 4564–4571 (2016)
33. Kim, D.Y., Kim, J., Kim, I., Jun, S.: Artificial landmark for vision-based slam of water pipe rehabilitation robot. In: International Conference on Ubiquitous Robots and Ambient Intelligence, pp. 444–446 (2015)
34. Balcilar, M., Yavuz, S., Amasyali, M.F., Uslu, E., Çakmak, F.: R-slam: resilient localization and mapping in challenging environments. Robot. Auton. Syst. **87**, 66–80 (2017)
35. Boiangiu, C.A., Bucur, I., Tigora, A.: The image binarization problem revisited: perspectives and approaches. J. Inf. Syst. Oper. Manag. **6**, 1–10 (2012)
36. Knuth, D.E.: Digital halftones by dot diffusion. ACM Trans. Graph. **6**, 245–273 (1987)
37. Bayer, B.E.: An optimum method for two-level rendition of continuous-tone pictures. In: IEEE International Conference on Communications, vol. 26, pp. 11–15 (1973)
38. Jarvis, J.F., Judice, C.N., Ninke, W.: A survey of techniques for the display of continuous tone pictures on bilevel displays. Comput. Graph. Image Process. **5**, 13–40 (1976)
39. Avola, D., Bernardi, M., Cinque, L., Foresti, G.L., Massaroni, C.: Adaptive bootstrapping management by keypoint clustering for background initialization. Pattern Recognit. Lett. **100**, 110–116 (2017)
40. Avola, D., Cinque, L., Foresti, G.L., Massaroni, C., Pannone, D.: A keypoint-based method for background modeling and foreground detection using a PTZ camera. Pattern Recognit. Lett. **96**, 96–105 (2017)

Author Index

Printed in the United States
By Bookmasters